A CULTURAL HISTORY OF INSECTS

VOLUME 6

A Cultural History of Insects
General Editor: Gene Kritsky

Volume 1
A Cultural History of Insects in Antiquity
Edited by Gene Kritsky

Volume 2
A Cultural History of Insects in the Medieval Age
Edited by Gene Kritsky

Volume 3
A Cultural History of Insects in the Renaissance
Edited by Sherilyn G.F. Smith

Volume 4
A Cultural History of Insects in the Age of Enlightenment
Edited by Matthew S. Lehnert

Volume 5
A Cultural History of Insects in the Age of Industry
Edited by Carol Anelli and Susan Fisher

Volume 6
A Cultural History of Insects in the Modern Age
Edited by Robert K.D. Peterson

A CULTURAL HISTORY OF INSECTS

IN THE MODERN AGE

VOLUME 6

Edited by Robert K.D. Peterson

BLOOMSBURY ACADEMIC
LONDON • NEW YORK • OXFORD • NEW DELHI • SYDNEY

BLOOMSBURY ACADEMIC
Bloomsbury Publishing Plc
50 Bedford Square, London, WC1B 3DP, UK
1385 Broadway, New York, NY 10018, USA
29 Earlsfort Terrace, Dublin 2, Ireland

BLOOMSBURY, BLOOMSBURY ACADEMIC and the Diana logo are
trademarks of Bloomsbury Publishing Plc

First published in Great Britain 2024

Series design by Raven Design
Cover image: Haematopota pluvialis,
the Common Horse Fly close-up © mikroman6/Getty Images

A catalogue record for this book is available from the British Library.

Library of Congress Cataloging-in-Publication Data

LCCN 2023029669 .

ISBN:	HB:	978-1-3500-0315-6
	Pack:	978-1-3500-0321-7
	ePDF:	978-1-3502-6669-8
	eBook:	978-1-3502-6668-1

Series: The Cultural Histories Series

Typeset by Integra Software Services Pvt. Ltd.
Printed and bound in Great Britain

CONTENTS

ILLUSTRATIONS

CONTRIBUTORS

Tierney R. Brosius is Associate Professor of Biology, Augustana College, Rock Island, Illinois, USA

Amelia Dolan is a science teacher at Heritage Community Charter School, Boise, Idaho, USA

David Kinkela is Associate Professor of History, SUNY-Fredonia, Fredonia, New York, USA

Barrett A. Klein is Assistant Professor of Biology, University of Wisconsin-La Crosse, La Crosse, Wisconsin, USA

Rayda K. Krell is Global Technical Education Scientist, Corteva Agriscience, Ridgefield, Connecticut, USA

Jeffrey A. Lockwood is Professor of Natural Sciences & Humanities, University of Wyoming, Laramie, Wyoming, USA

Robert K.D. Peterson is Professor of Entomology, Montana State University, Bozeman, Montana, USA

Gary Noel Ross is a Researcher, and retired Professor of Biology, Southern University, Baton Rouge, Louisiana, USA

Karl Zuelke teaches at the Art Academy of Cincinnati, Cincinnati, Ohio, USA

SERIES PREFACE

This six-volume set, *The Cultural History of Insects*, is a unique examination of how insects have infested the human experience. The set is divided into six time periods: Antiquity, the Medieval Age, the Renaissance, the Age of Enlightenment, the Age of Industry, and the Modern Age. Within each time period, eight topics are addressed: the state of entomological knowledge during the period; insects and disease; insects and food; insect products; insects' role in mythology and religion; insects as symbols; the portrayal of insects in literature and language; and the representation of insects in art. This structure is meant to allow readers to tailor their reading to their specific interests: for example, if insects' influence in the history of art is of primary interest, the reader can select the art chapter from each volume; if a reader wants to understand the impact of insects on a particular period in human history, then reading the volume on that time period would be appropriate.

The coverage of the topics presented is not encyclopedic. Presented are histories and case studies describing how insects influenced human culture and survival. The geographic focus is generally the Western Hemisphere, but details are included of geographic intersections in which cross-cultural influence plays an important role in various topics and time periods.

Insects were here before humans. They have influenced our evolution, eaten our food, provided us with raw materials, changed our history, and been the muses for our creativity. These six volumes illustrate how insects have helped to shape our culture over the past five thousand years.

Gene Kritsky,
Mount St. Joseph University,
Cincinnati, OH

Introduction

ROBERT K.D. PETERSON

The modern age is arguably the age of science. The scientific revolution that began in earnest a few decades before the period covered in this volume has profoundly changed humanity and human culture. In his book *Sapiens*, historian Yuval Noah Harari argues that the steep trajectory of progress wrought by the scientific revolution is fundamentally different than any other period in human history. Given that, what room is there for insects in modern *culture*?

Surprisingly, a lot. The cultural use of insects has evolved in the modern age. Whereas the use of insects in religion and the supernatural have subsided with the overall attenuation of religion in the modern age, insects are still used as good and bad cultural symbols in literature, language, film, music, and art. Moreover, knowledge about insects is used in science—especially as it relates to food, health, and products—and that knowledge is used culturally as well.

THE TRIUMPH OF SCIENCE: HEALTH, FOOD, AND INSECT KNOWLEDGE

Most of what we know about insects has come to us in the modern age through the practice of science. This is especially true when considering how we deal with insects both as pests and as beneficial partners (see Chapter 1). Through at least the first half of the twentieth century, insects were often seen as impediments to progress and worthy of absolute conquest. The ability to cure many infectious diseases as a result of the scientific elaboration of germ theory coupled with the discovery of effective insecticides led to a level of optimism that insect pests *could* be conquered (see Chapters 1–3). Indeed, cultural language in the first half of the century is replete with warlike terms and phrases when describing insects—and entomologists were soldiers on the front line (see Chapter 1).

The martial stance against insects was not without reason. Since before the dawn of agriculture, people were vexed by insects and they exacted a tremendous toll. Although no one understood the link between insects and disease, disease felled countless people. With agriculture came direct evidence of the burden presented by insects. People could watch their crops being destroyed and could see what was feeding on them.

The industrial and modern ages ushered in an optimism fueled by the power of science and technology to cure societal ills and improve the human condition. The study of insects consequently evolved from naturalism to experimentalism. Entomology became professionalized and bureaucratic and used the scientific method to advance (see Chapter 3). With that advance came the management of insect pests as a technological enterprise, aided greatly by chemistry. In addition to managing insect pests, using the scientific method on insects revealed fundamental objective knowledge about what insects are, how they evolved, how they live, how they die, and how they interact with other organisms in their environment. Moreover, using insects as study organisms opened up a world of new knowledge about genetics, physiology, molecular biology, evolution, and ecology.

The optimism provided by stunning scientific and technological achievements in the first half of the century bore out when it came to managing insect pests. Dramatic declines in populations of some pests as a result of using new synthetic insecticides caused some to think that pests could be eradicated, not merely controlled. The 1952 book, *Insects: The Yearbook of Agriculture*, published by the US Department of Agriculture, essentially argued that the war on insects was almost won. The optimistic argument was clearly influenced by the initial results in public health and agriculture from using the then relatively new insecticide DDT. The optimism became part of culture as DDT became common in household and garden products and thus became a common word with positive connotations (see Chapter 1).

The optimism was not to last. Despite the marvels of modern science and technology, resistance by insect pests became evident as soon as the technologies—especially insecticides—were used on a large scale. Resistance, notably genetic resistance, demonstrated clearly that idealistic visions of vanquishing insect pests and taming nature could not be achieved. In addition to resistance and resurgence by pests, new control technologies resulted in replacement of target pests with other pests and contamination of environments.

Problems with resistance and environmental contamination precipitated the publication of the landmark book *Silent Spring*, by Rachel Carson. This, in turn, led directly to the environmental movement, arguably culminating in the formation of the US Environmental Protection Agency in 1972. Cultural attitudes about DDT and pesticides in general went from positive to highly negative in the space of a single decade. Ironically, science had a direct role in dealing with the environmental problems that it helped create.

New scientific knowledge allowed for improved and more sustainable forms of dealing with insect pests. Of particular note is Integrated Pest Management (IPM), a "comprehensive approach to managing host stress that is economically and ecologically sustainable" (Peterson et al. 2018). From the 1970s to present, IPM has been the pre-eminent framework for managing pests and has even become part of cultural vocabulary (see Chapter 1).

The environmental movement also ushered an evolution in how culture views insects. Insects began to be recognized as integral parts of the environment, organisms to be respected—even cherished—for their roles in ensuring environmental health. There was greater recognition of beneficial insects, culminating in the current mass popularity of pollinating insects, such as bees and butterflies. Scientific knowledge of pollinators and other beneficial insects has contributed greatly to the current cultural appreciation of insects.

THE CULTURAL EVOLUTION OF INSECT PRODUCTS

At the beginning of the modern age, many traditional insect products were well known for millennia. These included honey, beeswax, silk, and dyes. There were also new insect products at the beginning of the modern age, such as phonographic records made with shellac, the resin made from the female lac bug.

A key event in the twentieth and early twenty-first centuries, however, has been the evolution in cultural thought about what a product is in general and what an insect product is in particular. Products have evolved from simple tangible commodities to concepts (see Chapter 4). This is also true of insect products. Today, insects are popularly known as providing products to people such as "ecosystem services." Ecosystem services are the benefits people obtain from specific environments and ecosystems. Examples of services include food, water, timber, flood prevention, water quality, recreation, aesthetics, photosynthesis, and soil formation. As the most abundant and species diverse organisms on earth, insects provide ecosystem services through pollination, seed dispersal, nutrient supply, dung burial, predation, and parasitism (see Chapter 4).

Pollination is arguably the most popular contemporary example of an insect service viewed by the public as an important environmental product. As a result of the environmental movement of the latter half of the twentieth century and contemporary media coverage of the plight of pollinators, bees have emerged as a cultural touchstone for the economic, aesthetic, and spiritual value of the environment to humanity. Bees are now thought of as products, not solely because of honey and beeswax, but also because of their value to ecosystems (see Chapter 4).

MODERNITY AND INSECT SYMBOLISM

As a key attribute of the modern age, science has caused insects as symbols to evolve in their use and meaning. Largely gone are insects as portending the supernatural and as omens for good or bad events. Insects now symbolize aspects of the science and technology that pervade the twentieth and twenty-first centuries.

Pollinators, such as bees and butterflies, now symbolize healthy ecosystems and natural harmony. The bee has become a major cultural symbol of environmental health, and "save the bees" is a common early twenty-first century phrase in activism and marketing campaigns (see Chapter 6).

Insects are also contemporary negative symbols of runaway science and technology. The apogee of this was the giant insect in film and literature of the 1950s and 1960s. These giant insects were often symbols of a curious mixture of runaway technologies and fears of communism (see Chapter 6). Public fear and fascination after the Second World War with radiation, nuclear power, and nuclear war expressed as giant insects that had to be destroyed. Ironically, scientists often came to the rescue to destroy those menacing insects.

In addition to science fiction literature, other genres of literature in the modern age contain numerous instances of insects as symbols. Newspaper columnist Don Marquis's character Archy the cockroach wrote comic poetry that was "emblematic of the contradictions inherent in literary and artistic Modernism" (see Chapter 7). Archy the cockroach was far removed from Romanticism's use of insects as symbols of idyllic nature—as was, most famously, Gregor Samsa in Franz Kafka's masterpiece novella, *The Metamorphosis*.

Gregor Samsa turns into a giant beetle (although some confuse it with a cockroach) and struggles to control his new body, even though his mind is fully human. Gregor's self-loathing and burdensome family caused him to metamorphose into this repellent insect. In modernist literary tradition, there is no deeper meaning here. Gregor is merely defeated and disintegrates over time in response to anxiety and uncertainty (see Chapter 7). In modernist tradition, life and death merely are. The fleas in Albert Camus's novel *The Plague* symbolize the relentlessness and inevitability of death—death without design or purpose.

All modern age literature, however, was not nihilistic existential angst. Vladimir Nabokov, the novelist *and* lepidopterist, used the butterfly in his writings as symbols of positive transformation, rebirth, and even spiritualism. In the novel *The Secret Life of Bees*, Sue Monk Kidd uses honey bees as symbols of the solidarity of women (see Chapter 7). Transformation was not always positive. The serial killer Buffalo Bill in Thomas Harris's novel (and subsequent blockbuster film) *The Silence of the Lambs* raises the death's-head hawkmoth as reinforcement of the power and perfection of metamorphosis. He wants to become a woman, but was denied a sex change, so he murders women and uses their skin as a cocoon from which he will transform into a woman—transforming into perfection.

Films in the late twentieth and early twenty-first century also expressed positive aspects of insects, especially as symbols of natural harmony. Although films such as *A Bug's Life*, *Antz*, *The Ant Bully*, *Bee Movie*, *Minuscule: Valley of the Lost Ants*, and YouTube's viral videos of *Lucas the Spider* have heavily anthropomorphized characters, they contain positive messages about insects and the environment.

In art during the modern age, insects have been used as symbols for many aspects of culture and the human condition, especially less than positive aspects. Examples include insects as symbols of climate change, habitat destruction, pollution, alien invasive species, species extermination, and bioengineering (see Chapter 8). Positive examples include insects as symbols of natural beauty, harmony, rebirth, and renewal. Not only are insects symbols in art, but they also *are* art. Insect exoskeletons and body parts have long been used as art (Marley 2008).

As with literature, film, and art, popular music in the modern age has witnessed numerous examples of insects as symbols. In early blues music, insects have served as animal tricksters, cunning characters in African narrative traditions that use humor and deceit for personal gain. Charley Patton's boll weevil, Blind Lemon Jefferson's mosquito, Furry Lewis's bed bug are trickster protagonists or antagonists, using their wits to get the best of their situation (see Chapter 6). Most insects in blues music are symbols for sex and sexual prowess. Bo Carter's song "I'm an Old Bumble Bee" is a prime example:

> I am an old bumblebee
> A stinger just as long as my arm
> I stings every good-looking woman now
> Everywhere I goes along

Bees and wasps are common subjects in blues music. Other examples include "Yellow Jacket," "Bumble Bee," "Bumble Bee Blues," "New Bumble Bee," "Honey Bee," "I'm a King Bee," "Queen Bee," "Yellow Bee," "Little Queen Bee," "Back Biting Bee Blues," "Little Bee," and "Stinging Bee Blues." A powerful example of the link between blues and rock music is Muddy Waters's song "I'm a King Bee." It has been covered by numerous rock bands, including tremendously popular groups such as The Rolling Stones, The Doors, Pink Floyd, and The Grateful Dead.

Insects serve as symbols of the heavy environmental footprint of human activity in songs such as "Maggot Brain" by Funkadelic, "Dragonfly" by Ziggy Marley, and "The World is a Dying Insect" by Kataklysm. Insects serve as metamorphic symbols of profound personal change in songs such as "Cuckoo Cocoon" by Genesis, "Moths" by Jethro Tull, and "Chrysalis" by Opeth (see Chapter 6). Butterflies are a constant source of inspiration with songs and album titles about the subject by Aerosmith, the Bee Gees, Cher, Heart, Mariah Carey, Smashing Pumpkins, and Widespread Panic (Coelho 2000).

The most common example of insects as symbols in rock music arguably is of human weakness and addiction. Songs such as "Moth" by Audioslave, "Bug" by Wussy, "Earwig" by Tenement, "Locust" by Machine Head, "Termites" by Tall Friend, and "Red Mosquito" by Pearl Jam use specific kinds of insects to dramatically expose the weakness, dependence, self-loathing, frailty, and regret of the musician and songwriter. The song "Bug" by Wussy, released in 2014, is a graphic example: "You're the drink, you're the drug | You're the bug that's alive inside of me."

Insects, of course, do not have to be symbols for anything in popular music. They can be subjects of fascination about nature, such as in "The Dragonfly" by Clutch. They can also inspire frenetic instrumentals, such as "7 Day Cicada" by LITE and "Ingesting Blattaria" by Blotted Science. Insects inspired musician names such as Bo Weavil Jackson and Bumble Bee Slim, and band names such as Buddy Holly and the Crickets, and—indirectly—the Beatles (see Chapter 6).

Modern English language has also imbued insect terms with other meanings. In contemporary culture, computers are literally an arm's length away from a sizeable portion of the earth's population. Because of that, nearly everyone knows what a computer or software "bug" is, although most probably do not know its link to real insects. Its use began in engineering in the 1870s. The use of the term to describe a glitch was derived from "bugaboo," which means a monster. The term's direct relationship to an insect is famously recorded in a log book now at the Smithsonian National Museum of American History. The computer engineer, Grace Hopper, in 1947 noted that an actual moth trapped in a relay created an error in a computer. The moth was removed and taped to the log book with the note, "First actual case of bug being found" (see Chapter 6).

The verbs "bug," "bugged," and "bugging" are used to mean either to annoy someone or to conceal a microphone to eavesdrop on someone. Bug as a verb has been used only relatively recently, increasing exponentially from 1960 to 1980, but then leveling off (Google Books Ngram 2018).

Some words that seem directly linked to insects are not. The noun "jitterbug" describes a dance popular in the US first in the 1930s and then around the world in the 1940s. In 1934, Cab Calloway recorded the song, "Call of the Jitter Bug." However, the lyrics of the song, the song title, and the dance have nothing to do with insects. Instead, they all refer to what happens when one drinks too much gin.

The word "shellacking" is an informal American term for decisively defeating someone or something. Its use in this way peaked in the mid-1940s, but it is still used (Google Books Ngram 2018). The formal verb means to varnish something with shellac, the resin shellac produced from the female lac bug.

The word "nitpicking" seems like it should predate the modern age. The activity of picking nits (lice eggs) from hair is older than humans and is a common practice of our great ape relatives. The word "nit" from the Old English "hnitu" has a much older Indo-European root, but, incredibly, "nitpicking" was not used until 1951 (*Oxford English Dictionary* 1989). It refers to fussy and pedantic fault-finding and it is not an admirable trait. How a term for an ancient act of grooming—an act that also promoted social cohesion—came to be used in the early 1950s to describe criticizing someone over unimportant faults is a mystery. Use of the term grew exponentially between 1960 and 1990, but has since plateaued (Google Books Ngram 2018).

The verbs "mothball," "mothballed," and "mothballing" have meaning in the twentieth century for cancelling or postponing work on a project, or long-term storing of equipment for potential later use. The words have been used in this way since about 1940, peaking in 1990.

***

The modern age contains the full spectrum of cultural attitudes about insects. For the first time in history, the technological ability to control insect populations led to a reinforced a view that insects are the enemy that needs to be feared, fought, and dominated (see Chapters 1 and 3). Insects prevented human progress and therefore were to be conquered—and scientific technology would lead the way to that conquest. There was—and still is—considerable loathing of insects (Lockwood 2013) and many people only see insects as disgusting little creatures that deserve to be killed.

Modernist existential angst produced Franz Kafka's Gregor Samsa as a beetle, Don Marquis's Archy the cockroach, and Pearl Jam's "Red Mosquito," but culture in the modern age also produced insects that harkened back to Romanticism and the Victorian nature movement. Vladimir Nabokov's butterflies, Sue Monk Kidd's honey bees, and E.B. White's Charlotte the spider point to positive transformation and a kinder, gentler reality.

Insects are cultural characters of great significance in the modern age. Indeed, they have arguably grown in importance from about the 1950s to the present as powerful symbols of human harm to the environment. Giant, mutated insects in science fiction books and films symbolize human-caused environmental disruption. Normal-sized insects at the dawn of the environmental movement to the present day are seen as symbols of natural beauty, harmony, and as bellwethers of something amiss. The emergence of the "save the bees" movement over the past decade has served as a remarkable example of the cultural embrace of insects, especially pollinating insects such as bees and butterflies. People are willing to use seemingly alien and hard-to-see creatures in contemporary culture as symbols of an environment—indeed, of the entire earth—that need to be protected.

CHAPTER ONE

Insect Knowledge

The Impact of Insect Pest Coverage in the Press on Cultural Attitudes about Insect Management

RAYDA K. KRELL

> The public must be made to appreciate the fact that it is a vital necessity to the future of humanity that we should learn everything about insects ... They must learn that the study of no aspect of insect life is a trivial pursuit, as their fathers thought and they must learn that the entomologists are the men [sic] on whom the world must depend for much of its future prosperity.
>
> (Howard 1925)

Leland O. Howard was Chief of the US Bureau of Entomology from 1894–1927 (officially deemed a "Bureau" in 1904) (Russell 1978) and the epigraph was part of an address he made to the Entomological Society of America in 1925. He highlighted the importance of bringing the work of entomologists into the public eye and suggested that the press was a perfect mechanism for gaining this publicity:

> This education of the public is going on rapidly, but we must speed it up. A number of strong writers in the newspapers and magazines have found the topic attractive and interesting. Many of us have grasped opportunities to address lay audiences of all kinds on the general subject, and all of us who can speak more or less forcefully in public must never lose a chance to drive this particular lesson home.
>
> (Howard 1925)

FIGURE 1.1 Leland Ossian Howard. Process print, 1911. Wellcome Collection.

Howard became a frequent voice in the *New York Times* in the early part of the era. Howard's many public statements were summarized as: "... Howard strove to boost public appreciation of entomology and to increase funding by framing insect problems as economic and public health issues" (Russell 1999). Later, there were critics of this approach that seemed to focus too heavily on the pest status of insects vs. a general appreciation (Russell 1999), but Howard's efforts were productive, and funding for the Bureau of Entomology increased during his tenure from about $30,000 in 1894 to $3 million in 1927 (Russell 1978). Also, the average human's interaction with insects involved their role as pests, so Howard's words rang true.

The cultural impact of insect knowledge in the modern era is dominated by the relationship between humans and the increased availability of more abundant and diverse tactics for pest management than ever before. The modern era began with relatively few insecticides such as rotenone and lead arsenate and has seen the addition of more than fifty insecticide subgroups (Sparks and Nauen 2015). The primary non-insecticidal tactics early in the era included cultural farming practices, but by the end of the era tactics that had never existed such as crops that are genetically engineered to resist pests and techniques to manipulate insect genomes have diversified potential options.

The impact of insect management on culture could seem difficult to measure, but we have a useful device because topics of human interest "make the news."

Presumably, anything printed in a newspaper was deemed worthy of public attention at the time of its publication. The concept of the press as a lens to view public knowledge of science has been used by other researchers (Evans et al. 1990; Clark and Illman 2006), but this lens has not been explicitly focused on insect management for understanding its perception across an era. *The New York Times* archive can serve as a single device to measure cultural attitudes towards insect management.

Howard's quotations set the tone for how entomology was perceived in the first half of the era through his passionate advocacy for the profession and its potential to benefit humanity. Howard keenly understood the power of gaining public attention in the press through engaging rhetoric and he was featured frequently in the paper in interviews and articles he authored (Howard 1926, 1929), creating an early-era high regard for entomology.

1920S: ENTOMOLOGISTS AS HEROES

Early in the era, the norm of cultural praise for entomologists in the press is easy to find, such as "... economic entomologists – the generals on the front lines of the battle of Man against Bug, upon the result of which our very existence depends ..." (Fisher 1929). Medical entomology had become an important part of the Public Health Service in 1920 with work on diseases such as malaria and Rocky Mountain spotted fever (Hughes 1961). The direct connection between insect control and human health was still fresh because it had been only been at the turn of the century that yellow fever had decreased from over twenty-five thousand cases to just twenty-six cases in a span of seventeen years through mosquito control efforts (Hughes 1961). Human plague and typhus were also brought under control by the mid-1920s through flea and louse control efforts, respectively (Hughes 1961). Respect for these achievements was common in the press (Anonymous 1922, 1923a, 1929).

The efforts to protect human health were impactful, but the relatively new ability to manage agricultural pests at a large scale would become a major theme in entomology coverage. In the early part of this era, insects were controlled by available insecticides, but most were inorganic and included ingredients such as arsenic, fluorine, copper, lead, and mercury (Hoskins et al. 1939). In 1921, new technology for insecticide application was reported when the first aerial application of an insecticide, lead arsenate, was used for catalpa sphinx moth over a six-acre forested area in Ohio. The historic insecticide application was publicized with the title "Hunting Bugs from an Airplane" (Anonymous 1923b). The article noted the potential of the new application method for protecting crops, which was an understated foreshadow of the later widespread use of this technology.

Walter Wellhouse, an entomologist at Iowa State College, commented on the new pest management trends: "With the growth of scientific research, men are gradually being able to control some of the natural forces which affect their welfare" (Wellhouse 1926: 2). A *New York Times* review of Wellhouse's book begins, "The enemy from which man has most to fear, the wise men tell us, is the insect" (Anonymous 1926b). The repeated common understanding of insects as enemies naturally drove cultural perception of entomologists as heroes.

It is likely that Howard's prominent quotes inspired thoughts, such as Wellhouse's. Howard reiterated a recurring theme about the need to manage pests and provide adequate resources, including trained personnel, for the battle. In a 1921 article, Howard cited government estimates that insects cause over $1 billion in crop damage each year (Anonymous 1921a) and highlighted concerns about the role of invasive species in these losses. Later that year, as president of the American Association for the Advancement of Science he feared that most people did not realize the critical situation of "... the enormous forces of injurious insects which attack him from every point and which constitute his greatest rival in the control of nature" (Anonymous 1921b). Howard's vivid language likely fueled his frequent appearance in the press and public interest in pest management.

In 1926, Howard delivered an address at the Pacific Division of the American Association for the Advancement of Science and said, rather dramatically, "If human beings are to continue to exist they must first gain mastery over insects" (Anonymous 1926a). The coverage of the address in the paper was followed by a full-page article he authored on the topic in the *New York Times* Sunday edition on pest control, titled "The Insects Intensify Their War on Man" (Howard 1926).

Howard detailed fundamental insect ecology with engaging language. He described many features of pest outbreaks, and, essentially, laid out the foundation of the practices that would much later be known and marketed as integrated pest management (IPM), even though he is not generally cited as a generator of the concept. He noted that when there is a pest outbreak, the primary goal is "... to bring about relief from great damage as quickly as possible; and the result is that chemical and mechanical means of destroying insects are likely to be first exploited." He even tackled the difficult task of explaining to the public how entomological research is critical to pest management:

> The investigators are carrying on all the time work on the biology of insect pests – studying their intimate life histories under all possible conditions – and it often happens that some weak point in the life round is discovered which indicates a cheap and successful method of attack.
>
> (Howard 1926)

He adeptly recognized that both cropping methods and introduced species often were the cause of pest outbreaks, but an understanding of basic biology and ecology was critical for targeting management tactics.

Howard made the potentially dry topic of insect management colorful with his descriptions of alternative methods of pest control, such as biological control: "Man has devices of his own for carrying on war with these lilliputians [*sic*] – gases and sprays and the like – but increasingly, he is obliged to employ insect mercenaries to fight for him" (Anonymous 1926c). He also discussed the potential for cultural control methods, such as those for the boll weevil and the European corn borer. He explained how an early-maturing cotton variety followed by destruction of crop residue held promise to manage boll weevils, but the practice was not being adopted by farmers. Instead, just a few years after the first report of an aerial insecticide application, airplane spraying of lead arsenate had become the preferred method of control. He described a cultural control method for the European corn borer that involved cutting stalks at the base to destroy overwintering larvae. But, he was not optimistic about the implementation of the method by farmers. At the time, there was concern about the spread of this invasive species to the corn-growing areas of the Midwest. Eventually, the European corn borer did spread and has remained a persistent problem for farmers throughout the country. Perhaps if Howard's early suggestions had been widely implemented as this early stage in pest invasion, the extensive geographic range and persistence of this insect could have been prevented.

Howard repeatedly made a case for why basic biological knowledge about insects was important for pest management: "What we really need is information on all the fundamental facts of insect life—their vital processes, physiology, and every aspect of their life from chromosomes up to the last reaction of the adult" (Howard 1926). He called for research institutions devoted to basic study of insects: "A research laboratory for the investigation of the fundamental things in entomology is a great need." The recognition that broad entomological understanding is foundational for pest management has ultimately become understood, but the need for more research institutions devoted to entomology remains on the wish list for entomologists today.

Howard's emphasis on the theme of human society being at risk from insects did not waiver throughout the decade and he made it the topic of his address to the 4th International Congress of Entomology: "This is not the age of man. It is the age of insects" (Anonymous 1928). He was relentless in his presentation of insects in this manner, which attests to his skill as an administrative leader in trying to leverage public opinion to ensure strong funding for the US Bureau of Entomology. In retrospect, Howard's engaging, but scientifically accurate, explanation of entomology and pest management to a general audience is impressive.

1930S: THE CUSP OF INTEGRATED PEST MANAGEMENT

Howard's dominant presence in the press continued with news of his book, *The Insect Menace* (Anonymous 1931b). Once again, he passionately conveyed that humanity was at risk for survival in its war with insects. As in earlier articles, practices he promoted were fundamental to the later formalization of IPM: "Dr. Howard urges farmers to change their methods of crop growing and make greater use of defensive weapons such as chemicals, and airplanes in the fight against insects." Thus, encouraging the use of both cultural and chemical, i.e., integrated, tactics. The book was reviewed a couple of weeks later with the headline "The Sinister Insect" (Kelly 1931). The article ends with high praise for a book on economic entomology intended for a general audience:

> It is impossible to give in the space of a review more than a suggestion of the varied riches of Dr. Howard's book. His half a century of study of insects has stored his mind with a marvelous treasure and this book is for the layman a fresh wonderland.
>
> (Kelly 1931)

The review seems to reflect a general high public regard for Howard and economic entomologists.

Howard was prolific in this period and published an additional book *Fighting the Insects: The Story of an Entomologist, Telling of the Life and Experience of the Writer*. This book was also glowingly reviewed in the press:

> For no medieval knight ever couched lance against a more formidable foe and no policeman ever captured a gunman more menacing to society. Certainly this battle against insects has its own heroisms, romances and thrills, and the reader of Dr. Howard's life story will find plenty of them in his pages.
>
> (Kelly 1933)

Once again, it seems the esteem of entomologists was largely fueled by Howard as a public figure and his stories of heroic entomologists saving various industries including citrus (twice), sugarcane in Hawaii, coconut in Fiji, and silk in Italy.

What is striking about this period is that strong concerns about the use of insecticides were already apparent, by both entomologists and the public. For the most part, by the end of this period, the tenants of what would become IPM existed; but the seductive efficacy of DDT that was on the horizon would be too tempting for most entomologists and farmers to ignore and would delay widespread adoption of an ecological approach to pest management.

The variety of methods available to control insects at the time were exemplified by an article by C.C. Hamilton intended for home gardeners, "Combating Insect Pests" (Hamilton 1934). Hamilton instructed: "Although the number of remedies and controls for these obnoxious enemies of the plant world is constantly being added to, plant growers have learned that there is no magic 'cure-all.'" He recognized what is, essentially, the first step in any pest management program, identifying the pest: "The first of these principles is that the 'cure' must fit the type of insect to which it is applied." He explained that different types of insecticides are appropriate for different pests and suggested stomach poisons, such as arsenate of lead and fluorine compounds for chewing insects and contact poisons such as nicotine, pyrethrum, and rotenone for sucking pests. He discussed insecticide mixing directions and following label instructions, which are all key concepts in modern instruction on insecticide application. He also mentioned the importance of good sanitation practices and keeping gardens relatively clean of plant debris to prevent providing additional habitat for pests. The scope of his suggestions remain tactics for pest managements practice today.

A milestone in biological control, the practice of deploying insects to manage pests, was picked up by the press. The first, and generally classic, example of using an insect as a biological control agent for a weed was reported in 1935. The example highlighted the beneficial role of insects as opposed to the focus on pests. The invasive prickly pear cactus had taken over thousands of acres of land in Australia. Many potential insect biocontrol agents were imported to feed on the cactus, but the Argentine moth proved effective because the larvae bore into the plant, efficiently killing the plant, which ultimately allowed for the reclamation of over fifty million acres of land (DeBach et al. 1976: 270). The impressive insect achievement was documented at least twice in that year: "... the district ... was choked by the useless prickly pear. Attempts to eradicate the pest were unsuccessful until the introduction of colonies of Cactoblastis [*sic*]. Now, not ten years since the insects were liberated on the pear, scarcely any of it remains" (Anonymous 1935a), and "The insect ended the appalling advance of the pear which was depriving the state of a million acres yearly" (Anonymous 1935b).

Biological control gained popularity as a topic in this period. The practice was described as "Two armies of shock troops totaling 15,000,000 and 16,000,000 insect parasites are being mobilized .... the parasite drafts are being called up for the 1931 offensive" (Anonymous 1931a). The article explains the challenges of rearing natural enemies for biological control deployment, but also the potential for success, which persist as issues for contemporary biological control programs.

In the 1930s air travel, including international travel, was increasing, and with it, an acknowledgment that insects could be easily transported between states and countries. "With great clippers of the air quickening communication between many foreign countries and the United States, the job of keeping out alien insect pests is bigger today than ever before for the Federal foreign-plant quarantine force" (Anonymous 1938). Once again, a problem identified early in the century remains a problem for the US Animal and Plant Health Inspection Service today.

It is appropriate to take a pause from newspaper reports and remember that heavy insecticide use was common for pest control and the entomological community was becoming more alarmed. In 1939, before the insecticidal properties of DDT were broadly known to the world, University of California entomologist William Hoskins and his colleagues wrote a paper titled "Recommendations for a More Discriminating Use of Insecticides," and essentially predicted what would come much later with the eventual formalization of IPM. Hoskins et al. wrote:

> ... the last few years [meaning the years before 1939 after widespread concern about the failure of inorganic chemical insecticides] may be considered as the beginning of an undoubtedly long period in which biological and chemical control are considered as supplementary to one another or as the two edges of the same sword which is raised for the protection of man's food from his insect rivals.
>
> (Hoskins et al. 1939)

He continued, "... it is instructive to adopt the viewpoint that all artificial control actually is aimed only at the portion of the population of a given pest which escapes its natural enemies." This description of population dynamics has become a tenant of modern courses on insect ecology. They described an approach to achieving this goal of more careful use of insecticides, "... only through accurate knowledge of the life histories, rates of development, effects of temperature and humidity, and of other ecological and physiological data is it possible to use chemicals in a truly scientific manner." This acknowledgment of the need to understand insect population dynamics is a fundamental component of contemporary IPM programs and was something Howard had tried to convey to the press throughout the period of increased insecticide use.

In this period, we were very close to avoiding the reliance on insecticides that was renewed and excited by DDT. The essence of Rachel Carson's later pleas were well established in 1939 by Hoskins et al. In the closing statement of their paper, they stated, "... agriculture as a whole, and entomologists in particular, intend to vigorously continue their efforts to diminish or remove the inherent dangers from the use of poisonous insecticides ..." but they could not predict that the transition to more integrated approaches would be interrupted by a new explosion in insecticide development that would begin with the commercialization of DDT in 1944.

1940S: NEW INSECTICIDES AND OPTIMISM ABOUT INSECT CONTROL

In the early part of the decade, before widespread use of DDT, entomologists were beginning to notice more frequent insecticide failures, "In his AAEE [American Association of Economic Entomologists] presidential address, H.S. Smith (1941) in a calm, reasoned essay, pointed out that the selective pressure of insecticides doomed their effectiveness" (Smith 1989). Smith wrote:

> We can no longer say positively that the control of some particular insect is a solved problem. The solution may be only temporary ... the economic entomologist can expect to be continually faced with the prospect that many of his recommendations will, in time, become of doubtful value.
>
> (Smith 1941)

But, he also saw the potential to restore efficacy:

> The point is, that, in general as these populations shift from one race to another, we should be able to alternate the treatment in such a way as to cause them to shift back .... And so, the whole problem of "how insecticides kill" grows in importance, and expansion of research in insect physiology becomes a paramount necessity.
>
> (Smith 1941)

Smith recognized that the practice of insect pest management requires the coordination of many disciplines. Smith also understood that while economic entomology was an applied science, the observations of insecticide resistance could contribute to "the nature and methods of evolution." This conversation that was going on among entomologists appears to have not been picked up by the press. It is possible that Howard's early precedent for rhetoric that popularized the importance of positive entomological research reduced the public appetite for caveats to that narrative.

This decade opened the new era of insecticides following the discovery of the insecticidal properties of DDT in 1939 and its subsequent use during the Second World War. "A new age in entomology began with World War II. Entomology emerged from its traditional obscurity and took its place as a full partner in the free world's struggle" (Smith 1989). This was partially because of the successful use of DDT by medical entomologists during the war and their perception as heroes for their role in protecting the military.

The decade was not entirely dominated by synthetic insecticides. It also brought new innovations in insect management such as one of the early successful uses of microbial control against a pest. In 1921, a disease that could kill the invasive insect, the Japanese beetle, had been recognized. Later it was determined the disease was caused by a bacterium that was later named *Bacillus papillae*, after the scientific name of the Japanese beetle, *Popillia japonica* (Lord 2005). The bacteria were eventually marketed and sold as "Milky Spore." The new control method and the patent for the insecticide were reported:

> … an insecticide which contains the spores of bacteria deadly to the larvae of the beetle. Feeding on the soil sprayed with such an insecticide, the larvae become infected with the bacteria which eventually kill the grub by the so-called milky disease which the bacteria produce.
>
> (Anonymous 1941)

Generally, the public would often be interested in such novel tactics, even though many of them would not have pervasive management success.

DDT became available on the American market in 1945, even to home gardeners. The excitement and promise by this new chemical was widespread: "… entomologists predict that DDT will have an important place in insect control along with other materials already in use" (Anonymous 1945b). It was certainly a correct prediction, but the scope of the impact was likely not part of the prediction.

Even though DDT had just come onto the market in the US in 1945, some entomologists expressed concerns about its environmental effects, including killing beneficial insect populations and toxicity to amphibians (Anelli et al. 2006). The concern was voiced by the press: "Further tests with DDT must be conducted to reduce its harmfulness to bird and animal life and determine the degree of concentration that must be used …" (Anonymous 1945b). The report discussed a National Audubon Society symposium in New York at which Audubon members, entomologists, federal officials, wildlife managers, and chemists all participated. The concerns were based on observations after application, but all seemed to acknowledge the efficacy and promise of the chemical for controlling pests. Dr. Curran, from the Department of Entomology at American Museum of Natural History (AMNH) said, "… it is perhaps the most valuable insecticide that has yet been developed," but he continued to say that because of the effects on aquatic life, the chemical would no longer be used at AMNH property in Bear Mountain and Harriman Park (Anonymous 1945c). Thus, twenty-seven years before DDT was banned in the US, some were already implementing limitations on its use.

Additionally, concerns about home applications were shared: "The United States Department of Agriculture has made studies of DDT to discover its potentialities and dangers. In the hands of the Army's trained personnel DDT was a brilliant success. Civilians need to be told how DDT should be used"

(Kaempffert 1945). It was acknowledged early on that DDT was toxic to nontarget organisms:

> DDT is harmful to honeybees and other beneficial insects, so that it may interfere with cross-pollination. It may destroy insect parasites and predators. It is highly poisonous to fish and some other cold-blooded animals. It has even killed birds that feed on insects.
>
> (Kaempffert 1945)

But, generally, the convenience outweighed these concerns. The general sentiment was summed up as "DDT is harmful to honeybees and other good insects, and to fish. A residue of it in the soil may retard the growth of some crops. In spite of these drawbacks, its future looks bright" (Anonymous 1945b).

The ability to simply make life more comfortable with DDT was reported: "A helicopter sprayed the Yale Bowl with insecticide prior to a recent Pop Concert so that music lovers could enjoy the program without mosquito-slapping" (Anonymous 1945a). Once again, despite early concerns about widespread DDT use, the advantages and comforts it could provide outweighed early reports of risks.

By 1947, DDT was considered, "a common word in every household however humble or remote" (Lyle 1947). Advertisements for household products with DDT were easy to find: "Bontox ends garden pest guesswork! This one low-cost dust frees gardens from both insect and fungus injury – nothing more to buy! Its Rotenone, DDT, New Type Copper, and other toxicants safely protect vegetables, flowers, shrubs, and berries, against most the insects and fungi known to invade gardens" (Anonymous 1947a).

With these early reviews of insecticide success, it is easy to perceive the public likely enjoyed a general sense of wellbeing and comfort that many of the insect problems in near memory would no longer be a nuisance.

But, in the background, entomologists continued to acknowledge concerns early in the expanded use of DDT about broad-spectrum effects and too much reliance on a single management method. Percy N. Annand (1947), who became Chief of the Bureau of Entomology and Plant Quarantine fourteen years after L.O. Howard, highlighted the importance of insecticides as tools, but cautioned:

> The enthusiasm with which these new tools for insect control have been received and the attraction in doing work with new materials promising quick results have, I fear, directed attention momentarily away from many of the other important though less spectacular phases of insect control. The thought that work on insect biology, physiology, ecology, taxonomy, and

> other basic problems and on cultural and biological control is of little or no importance now that giant killers have arrived may be persuasive to some.
>
> (Annand 1947)

The concerns raised about overreliance on a single technology have unfortunately been repeated in pest management each time a new technology is introduced. It is possible that Howard's early rhetoric to frame human interaction with insects as a battle for survival pervaded in human consciousness. Even though Howard's actual message was more nuanced with his explanation of the need to understand basic pest biology, the headlines about the threat to humanity caused by insects were likely more memorable and fueled prioritization of killing insects vs. studying them.

As early as 1947, the concerns about DDT accumulation in meat and milk were shared in the press: "Congress has found out that DDT shows up in the meat, milk and butter of cows fed on crops dusted with the insecticide" (Anonymous 1947b). The short article shared Percy Annand's short testimony in which he also noted the concern about DDT killing the parasites of insects that then allowed pests to flourish.

Despite some flashes of concern, the overwhelming opinion of the new prospects for pest control were still primarily exuberant. At an international congress on tropical medicine and malaria, a demonstration of pesticide application was presented:

> ... the latest types of insect killers, provided the most striking evidence presented so far to most of the foreign scientists that, for the first time in the million years man has lived on this planet, he had the means for gaining complete mastery over the insects that threatened his existence on two fronts – the carriers of deadly diseases and the destroyers of vital food crops.
>
> (Laurence 1948)

The new prospect of sanitizing new areas from insect pests, especially those of medical importance, was seen as way to alleviate many emerging global problems:

> ... areas in the tropics could be added to the habitable areas of the globe, thus greatly relieving the pressure of growing world populations and the growing scarcity of foods and raw materials. Most of these insect-infested regions of the world have hardly been explored for their mineral resources.
>
> (Laurence 1948)

The excitement led to optimistic, but ultimately false, predictions: "Within five years, they added, 'we will be able to control completely such major agricultural pests as the boll weevil'" (Laurence 1948).

New research into the potential of "feeding ... animal hosts with certain chemicals that will make the blood deadly" was reported (Kaempffert 1947). This early work was led by E.F. Knipling and colleagues at the Bureau of Entomology and Plant Quarantine in Florida (Knipling et al. 1948). The research reported feeding different insecticides to rabbits to determine if it could deter and kill pests such as lice, mosquitoes, ear mites, and ticks. This early work certainly supplied a leap in new ways to control invertebrate pests on domestic animals, a concept that is common today on pets and livestock.

In 1948, in light of the weight of publicly perceived positive contributions, Dr. Paul Mueller was awarded the Nobel Prize in Medicine for his discovery of the insecticidal properties of DDT (Anonymous 1948). He had actually first tested the chemical to manage Colorado potato beetle on Swiss potato crops, but the award acknowledged its broad use:

> A man who could save not only crops, at a time when we had to feed the world as well as our own forces and ourselves, but also the lives of hundreds of thousands of soldiers and civilians deserves the highest honor that science can bestow.
>
> (Anonymous 1948)

It is clear indication of the widespread positive perception of usefulness of insecticides at the time.

1950S: REPORTS OF INSECT RESISTANCE AND RENEWED INTEREST IN BASIC ENTOMOLOGY

In the early 1950s, insect resistance to DDT was already making the news: "Mosquitoes of the present generation are several times harder to kill with DDT than were their ancestors when the chemical was first used for that purpose" (Anonymous 1952). A report to the New Jersey Mosquito Control Association further explained: "... studies made in Florida and southern California had demonstrated that the longer a marsh area received annual DDT sprayings, the more resistance the adult mosquitos and their larvae showed." The results led to the recommendation to return to "... more permanent mosquito control work in marshes such as drainage and filling, and less dependence on chemicals." It had been less than ten years since the widespread adoption of DDT in the US for pest control and alternatives were being suggested. A call was also made in the article to test all insecticides on potential effects on wildlife.

Despite the suggestion that insecticides would not provide a sole solution to pest problems, the development and the search for more products continued and optimism about insecticides that would be impervious to resistance was reported: "A new insecticide called malathion ... will not produce resistant strains of insects ..." (Anonymous 1953a). It was considered one of the safest insecticides to handle and "The number of insect generations on which tests have been conducted to date far exceeds those on which D. D. T. and other chlorinated hydrocarbons can be applied without producing resistant strains." Of course, the lesson of insecticide resistance still had not been cemented. Behavioral and physiological resistance to malathion would be recorded within the decade in 1956 (Hammerstrom 1958).

The 1950s emerges as an era where the importance of understanding insect physiology, especially in the context of growing insecticide resistance, was gaining greater acknowledgment. An article on a report from the Entomological Society of America included details of new research on fly physiology:

> These and related researches have been brought about by the realization that DDT and other new compounds are not going to exterminate the housefly. Flies have become increasingly resistant to these chemicals ... One of the weak points in the battle against flies is the lack of knowledge about how the chemicals do their work. Little is known about housefly physiology.
>
> (Anonymous 1953b)

In a report from the next day of the same meeting, insecticides that could replace DDT, chlorthion and diazinon, were presented: "They are described as organic compounds to which house flies would not be able to develop resistance" (Anonymous 1953c). In discussing DDT it was called "a 'complete flop' on houseflies" and there was much hope about the new "systemics that attack insect pests inside the plants." Thus, despite prevalent concerns, insecticides were still seen as a positive solution for pest problems.

This decade ended with the beginning of what Chapman (2000) called "the most spectacular insect control event of the century," namely use of the sterile insect technique for eradication of the screwworm fly in 1959. The technique was theorized by Raymond Bushland and Edward Knipling based on the discovery that female flies mate only once in their lifetimes. They developed a technique to sterilize male flies and release male flies so that females would not produce viable offspring. Several years after implementation, the event gained attention in the news with the headline "Atomic Energy Being Used to Eradicate Screwworm" (AP 1962). The widespread success of the technique would require decades of dedicated focus, but would become a model for the value in understanding insect biology for non-insecticidal insect management success.

1960S: THE DOWNFALL OF DDT AND NEW INTEREST IN INSECT PHYSIOLOGY

The primary story of the 1960s involving entomologists was the publication of *Silent Spring*. The book would become to be known as one of the most influential books of the twentieth century, but the first report of its upcoming publication did not portend what was to come: “Rachel Carson ... has written a book about the effects of pesticides on the ‘balance of nature’” (Anonymous 1962). The second report of the book’s publishing seemed to better grasp the impending impact: “The $300,000,000 pesticides industry has been highly irritated by a quiet woman author whose previous works on science have been praised for the beauty and precision of the writing” (Lee 1962). Three articles with portions of the book had already appeared in the *New Yorker*, but their publication had not been met with the intense criticism that the impending publication of the book was receiving. The quotes in the article from pesticide industry employees imply they were hurt by her book’s implications: “... she has chosen to ignore the enormous benefits in increased food production and decreased incidence of disease that have accrued from the development and use of modern pesticides,” and “... we are quite concerned over the misrepresentation of an industry that has tried to do right.”

In the year following publication of *Silent Spring*, a nuanced review of the debate was published that balanced the concerns with the realities (Galton 1963). The problems of how to handle new invasive species and the public demand for perfect produce was presented. The article cited a National Academy of Sciences report that stated, “It seems indicated the American people cannot be fed adequately unless crops and livestock are protected from insects and other pests ... chemical pesticides are as much a part of farming as the tractor” (Galton 1963). There was an emphasis on how all chemicals can be toxic if improperly used. The article shared that there is general agreement that DDT does accumulate in tissues, but the effect of any accumulation was noted as a point of disagreement.

The article included several counter-assertions to Carson’s fundamental premise that insecticides upset “the balance of nature.” George Decker, a past-President of the Entomological Society of America in 1955, tried to put the use of pesticides in perspective: “It would be more accurate to say that pesticides are used to suppress organisms already out of balance.” And the *Progressive Farmer* magazine was quoted in the story:

> We upset the balance of nature to some extent whenever we break ground for a crop, kill a timberwolf or cut a tree to build a house .... It is only through disturbing nature's balance by intervening drastically in his own behalf that man has been able to live on earth in a civilized way.
>
> (Galton 1963)

Possible alternatives to pesticides were discussed such as cultural practices, resistant crop varieties, biological control, insect sterilization, as had shown promise with Knipling's work, and new hope based on work with insect hormones to use the products against pests. The consensus seemed to be that "It seems clear that pesticides have made important contributions to human health and progress but they can be dangerous when abused—and that they have been abused in the past" (Galton 1963).

It is interesting here to note that this discussion had been addressed in 1925 by L.O. Howard. He addressed the paradox between human needs, which upset the "balance," and how research into insect interactions should be understood and leveraged:

> Insects are controlled by nature very much better and more effectively than by man himself—assuming that man's efforts are not a part of nature (as they surely are, since man is a true ecological factor). Many more insects are destroyed by their own parasites of different kinds and by their other natural enemies than are destroyed by man. In fact, man has really been making things easy for the insects—he has been facilitating the multiplication and spread of his own worst enemies, and at the same time he has been groaning and complaining and suffering from the effects of his own stupidity.
>
> The trite words "going back to nature" have an especial significance in this connection. We cannot go back to nature, in a way, if we wish to reserve our rank as the dominant species and still keep on increasing at the time. We must continue to "upset the balance" and keep on upsetting it more and more if we are to feed our increasing millions. We must control all possible food sources. But, we must get back to nature in so far as we must study all these natural control factors as they affect antagonistic species, we must utilize them.
>
> (Howard 1925)

Howard addressed concerns about the balance of nature, and even called out how humans created conditions that turn benign insects into pests. But, he also took a practical stance by acknowledging human needs and then challenging researchers to understand and promote existing biological mechanisms of insect management whenever possible. His address had been for an entomological audience, and again, some of that nuance had not been as powerful as the more dramatic presentation to the public of the war with insects.

Insect physiology gained attention with a report of the 12th International Congress of Entomology which included an address from V.B. Wigglesworth, one of the key figures in insect physiology since the 1930s. He continued the emphasis on the promise of new understanding of insect physiology that would come from discoveries on insect hormones (Anonymous 1964). This potential was reported again on the work of Carroll Williams and colleagues that they had synthesized a juvenile-hormone-like chemical (O'Toole 1966). "... it appears to act so differently from poisons like DDT that it might open the way to the development of a new kind of insecticide that would be harmless to all animals that are of a higher form that insects." The primary drawback of the chemical that was noted is that it was active on all insects, even those considered beneficial, so the researchers noted they would like to develop something more selective. Generally, the public liked the idea of a "juvenile hormone" and just a few years earlier when Williams's work was reported the article speculated:

> Should the hormone turn out to serve as a 'Peter Pan' hormone in man, it will open a new vista for space travel to distant planets .... A space traveler equipped with the synthetic hormone would always remain at the same age at which he started on his journey.
>
> (Laurence 1958)

The name "juvenile hormone" seemed to grab attention.

The excitement about the diversity of new potential methods for insect management was well documented in an article highlighting the range of research into new methods. "In fields and laboratories throughout the country scientists are seducing insect pests with love potions, forcing them to practice birth control, jamming the radar they depend on for navigation, and fooling them into celibacy" (Brody 1967). The article highlighted the advances in pest management with the sterile insect technique and new work into using sex attractants (now commonly called pheromones), to confuse insects and disrupt mating or lure them into traps. The interest in juvenile hormone for pest control was again mentioned. Despite the excitement surrounding the potential of new tactics, the role of insecticides was acknowledged and perhaps one of the earliest mentions of "integrated control" in the press was mentioned: "... the formula for today ... is 'integrated control' – a combination of chemical, biological and other measures to obtain the most effective results with the least amount of chemical poison and the least hazard to man and wildlife."

1970S: RENEWED INTEREST IN PEST MANAGEMENT BASED ON INSECT ECOLOGY

The real impact of Carson's book on entomology would be felt in the early 1970s, especially when DDT's use was cancelled in the US in 1972. In an article on the future of pest control without DDT, the drama of the situation was described: "… the recent fall from favor of pesticides such as DDT – once an honored warrior in the battle against bugs but now a recognized killer of wildlife and perhaps even a threat to man" (Higdon 1970). The article focused on the suggestions of Edward Knipling, the Director of Entomology research for the Agricultural Research Service, for a vision of what pest management could be in the future. Knipling highlighted that when an insect is a "pest" nature is out of balance (in response to Carson's premise that insecticides interfered with "nature's balance." This concept had been noted earlier by George Decker, and even earlier by L.O. Howard). Knipling continued, "It's my view that we can eradicate an alien insect from our system and it won't affect the balance of nature. In fact, I argue that you are restoring the balance." He was interested to explore the sterile insect technique for more insect pests, and there was a new area of work looking into chemosterilants that would be sprayed on insects in the environment, vs. needing to rear, sterilize, and then release millions of insects. He also discussed the new potential of pest management based on understanding insect hormones and pheromones. Knipling framed pest management going forward as having an intent to be more selective: "When we talk of selective methods, whether they be biological or chemical, we mean controlling the given insect without harming other things." Knipling did not discount a role for chemical methods. The nuanced view of the complications regarding insect management was receiving more central coverage in the press.

The shift to pest management involving a variety of techniques was well explained in a 1976 article (Brody 1976). The article quotes many of the important entomologists who were active in spreading the popularization and efficacy of integrated pest management including Robert L. Metcalf, Carl B. Huffaker, and Dale Newsom. The article explained the relatively new concept of only spraying when pests occur at population sizes that can cause damage to crops: "… they spray 'only when needed' – that is, only when the pest is there in sufficient numbers and in the pest's life stage that can cause economic injury." The realization that pests could be tolerated at certain levels without causing economic damage was a new concept:

> Determining the "economic injury level" or how many of each pest must be present before the crop is significantly damaged … researchers figured out how much food a single insect of each pest species consumed in

> development. They concluded that much larger numbers of soybean pests could be tolerated than anyone had realized before costly damage would be done.
>
> (Brody 1976)

The use of computers to create pest warning systems based on weather data was discussed as one of the decision-making tools. In addition, the use of trained field scouts to monitor fields for pests was discussed as a new intervention as compared to past spray programs that were established on a preventive and scheduled basis. The article acknowledged that many of the concepts in pest management had been presented as early as the 1920s, but the extreme efficacy and convenience of the new pesticides, such as DDT, had been too miraculous to provide incentive for more a more integrated, and complicated approach.

As the decade closed, the prominence of IPM was front and center in politics:

> Integrated pest management, as Congress will learn this week, is more than a fancy name for swatting flies and spray-bombing ants. An innovative comprehensive approach, it combines crop ecology studies, biological controls, meteorological information and other disciplines and data to protect crops from bug infestation. And with very few pesticides.
>
> (Webster 1979a)

The array of practices that could be part of IPM were listed:

> To save a crop from its insect predators, it may be necessary to employ pheromones, the sex scents that lure male insects away from females; viruses or bacteria; juvenile hormones and growth regulators; judicious amounts of pesticides; predator insects; or disease and insect-resistant plants; and to combine appropriate techniques with knowledge of the plant's ecology, planting schedules, and weather patterns.
>
> (Webster 1979a)

Fans of the approach were found as high up in politics as the presidency: "... President Carter has called twice for Federal agencies to speed implementation of integrated pest management programs ...." Perhaps Carter's background in agriculture facilitated the promotion of IPM, but it was clear the idea was not hidden in the confines of land grant entomological discussions.

In 1979, IPM was mentioned in an article as a solution proposed by the United Nations to ameliorate the growing number of insects resistant to insecticides (Webster 1979b). The article included a graphic to illustrate the scope of methods that could be included in an IPM approach. The article accurately presented the problem and the concepts in IPM to a non-agricultural or non-scientific audience and IPM had become part of relatively common national vocabulary.

1980S AND 1990S: BIODIVERSITY AND BIOTECHNOLOGY

In the 1980s, there was growing appreciation of insects and their diversity, as the new concept of "biodiversity," credited to entomologist E.O. Wilson, was gaining popularity. He was quoted: "I'm astonished that something as basic and important as the amount of diversity of life on earth is largely unmapped" (Eckholm 1986). The article highlighted work by Terry Erwin in the tropics: "... [he] has developed a technique for fogging forest trees with insecticides, allowing systematic collection of all the insects that drop ... Dr. Erwin extrapolated the existence of at least 30 million species of insects." Before Erwin's research, there were only 1.6 million described species, with approximately half being insects, so Erwin's work had completely changed the understanding of insect diversity on the planet (Erwin 1982).

E.O. Wilson's more vocal role as an advocate for conservation during this period brought more public attention to his contributions to science and society. In an article profiling his work, his many books and work on ants was highlighted, as well as his more recent role as a defender of biodiversity (Royte 1990). Another lasting contribution to thinking on biodiversity was entomologist Quentin D. Wheeler's depiction of the number of species in different animal groups by use of a new graphic termed a "speciescape" (Anonymous 1990).

The graphic gained attention for the relative view of the importance of various biological groups based on their diversity. The concept and versions of the graphic have become a standard feature in most introductory biology classes.

Concurrently, IPM was becoming the prevailing philosophy of pest control: "... it is better to manage pests and limit them to tolerable levels, an approach known as integrated pest management or I.P.M." (Faust 1987). The article detailed the diverse methods available for pest management including pheromones and biological control. The article also discussed the potential for biotechnology to play a role in the future: "Biotechnology is also entering the field of pest control ... future controls will include some gene management of pest species."

Indeed, biotechnology became some of the biggest news in entomology and pest management (Feder 1996a) with the introduction of corn genetically engineered to resist the European corn borer (the same pest that

L.O. Howard had used as a case study in the 1930s). The image of "... eight shiny rows of corn that look as if they are right out of a seed company's sales brochures alternating with eight rows of stalks with leaves scarred by feasting insects" was the opening verbal visual in one article (Feder 1996b). The first genetically engineered corn with a gene that produced a Bt protein was approved that year. In the article, the lack of research into resistance management and the lack of basic biological knowledge about the insect pests was mentioned. It was noted that past reliance on pesticides had lessened interest in basic biological questions such as how far European corn borers fly:

> One problem for modelers, scientists say, is that until relatively recently, the actual biology and behavior of many important insect pests have received remarkably little attention. The focus in agriculture has been on the performance of pesticides and alternative controls like crop rotation.
>
> (Feder 1996b)

The need for sociology and psychology to understand why and how to encourage farmers to adopt recommended refuges was mentioned. The increased interest in looking to other disciplines to learn how to better communicate research to farmers continues to be a contemporary area of interest for entomologists.

The resilience and genetic diversity of the insects was again exemplified with reports of "rotation resistant" corn rootworms:

> In the endless battle between farmers and insects, the bugs have won another round. A big change in the behavior of the western corn rootworm beetle is threatening one of the great success stories in American agriculture, the use of crop rotation instead of pesticides on millions of acres of Midwestern corn.
>
> (Thomas 1998)

There was new hope about the development of a genetically engineered corn with a Bt-protein toxic to corn rootworms. In response to the hope for a rootworm cure, "... this would provide only temporary relief." Quoting Joseph Spencer, "It's another magic bullet .... We've fired them before. It may kill one insect, but a resistant variant may arise in its place. It's like a phoenix. In agriculture, problems are not solved forever." The statement reflected what the evidence of history made clear; pest management must constantly evolve as the pests evolve. This again proved to be true with the introduction of Bt corn resistant to corn rootworms in 2003 (Pollack 2003) and the first reports of rootworm resistance to the corn in 2011 (Gassman et al. 2011).

ENTOMOLOGY AT THE BEGINNING OF THE TWENTY-FIRST CENTURY

The story of insect pest management in the twenty-first century is just beginning, but a developing theme is the equal dual interest in managing pests and preserving insects for beneficial purposes and an appreciation of species diversity. The need to develop new pest management tactics continues with a continual increase in the introduction of invasive insect species and the ongoing development of pesticide resistance. Concerns for managed and native bees has been a focal point in the news with reports of bees dying (Johnson, K 2010). New genetic techniques offer increasingly innovative and specific tactics for controlling insects such as designing mosquitoes that produce flightless females making it impossible for them to mate (Koerth-Baker 2013) and mosquitoes incapable of carrying pathogens (Pollack 2016). Another new method termed RNA-interference, or RNAi, that is actively being explored as either a pesticide spray or for incorporation into genetically engineered plants. The technique:

> ... harnesses a Nobel Prize-winning discovery to kill insects and pathogens by disabling their genes. By zeroing in on genetic sequence unique to one species, the technique has the potential to kill a pest without harming beneficial insects. That would be a big advance over chemical pesticides.
>
> (Pollack 2014)

It is clear there will be new innovations in pest management, but the history of human-insect interactions shows that insects will challenge each new innovation. Engaging the public though platforms such as the press are important tools for fueling the public's interest in pest management and driving funding for research, but care must be taken to ensure the nuance of pest management makes the headlines and not just the drama. Past rhetoric referred to the human war with insects, but going forward, perhaps a better approach is to consider the relationship as a friendly rivalry, where each challenge is accepted as the opportunity to learn something new.

CHAPTER TWO

Insects and Disease

Insects, Disease, and War

JEFFREY A. LOCKWOOD

In a broad sense, we can think of there having been a series of revolutions in western civilization over the last five hundred years, including the Reformation in the 1500s, the Enlightenment in the 1700s, Romanticism in the 1800s, and Modernism in the 1900s. While this rather oversimplifies history, such an approach provides a useful heuristic. And so, the cultural history of entomology in the twentieth century can be understood within the framework of the dominant concepts of this period.

If Modernism is the zeitgeist of the twentieth century, then it makes sense to delve a bit more deeply into this historical framework (Childs 2017: 16–17). The roots of Modernity can be traced to the Renaissance as scholars sought a vocabulary to describe modern life. In the next phase, the American and French revolutions generated social upheavals that called into question old practices and catalyzed notions concerning the mastery of nature and society through the use of reason. And in the third phase modernization spread globally in the 1900s.

Using Modernism as the thematic structure for viewing entomology in the twentieth century risks terminological confusion. Most often, Modernism is applied to the arts, but for our purposes it can be taken as an enveloping term, albeit having many interesting points of contact with the fine arts that there is insufficient space to explore here. Modernism is also entangled with the ideas of Neo-Victorianism and the Enlightenment, along with Structuralism—and it also gives rise to the reactionary movements of Post-Modernism/Structuralism. And so for clarity, some explication of Modernism as applied to culture is in order.

MODERNISM: AN HISTORICAL FRAMEWORK

To make sense of how Modernism will be used in this analysis, consider the following five elements that will be woven into the cultural history of entomology: ideology, scientism, pragmatism, relativism, and corporatism.

The unifying concept of modernist *ideology* is that although there are changing alliances and allegiances, some collective grouping (e.g., race, gender, nation, government) is superior to the others. Foucault maintained that Modernity was the attitude of placing humanity and reason at the center of everything, so in this sense the superior entity is *Homo sapiens* (Childs 2017: 16–17). If so, the aspiration of this ideology is to emancipate all people. However, critics argue that reason and knowledge became the means for enslaving inferior groups, much as had been done in pre-modern societies through religion and social authority.

So it was that imperialism became a kind of autocatalytic ideology as European nations were exposed to alternative cultures and had to come to terms with what were often stark differences (Cantor 1988: 6; Childs 2017: 20). To justify colonization, the western cultures staked out a position of moral and social superiority.

Following the First World War, modernist ideologies were strengthened by reflecting on how dismally the old social order had failed (Cantor 1988: 132). At the same time, there was something of a revival of nineteenth-century culture, and this neo-Victorianism fed a fervent nationalism as people sought stable social structures in the wake of appalling human suffering.

Although modernist ideology can be manifest in terms of imperialism and nationalism, it is worth noting that Modernism did not generate a distinctive political ideology (Cantor 1988: 4). Rather, as seen in the early half of the twentieth century, various forms of chauvinism were manifest—including notions of white superiority (Childs 2017: 6)—which were not restricted to any particular political system. The key feature was a kind of social fragmentation and discord that underwrote an us-them distinction which developed into a better-worse dichotomy (Cantor 1988: 35–9).

After the Second World War, Structuralism shifted the cultural focus from the individual to the system, with the rise of multinational corporations. What this meant in ideological terms is that imperialism waned along with social Darwinism and eventually gave way to Deconstructionist and Postmodernist views (Cantor 1988: 345). So instead of making judgments about certain groups being superior to others, social relativism prevailed—a view that all people and practices are of equal value and judgments of better/worse merely reflect social preferences.

***

According to *scientism*, humans battle against nature and one another, with victory coming through the power of science and technology (Sorell 1991). There is an abiding faith that technological sophistication will allow certain cultures to wrest control in the midst of a chaotic and dangerous world.

This view was rooted in the Renaissance and Reformation but came to flourish with the Industrial Revolution and the emergence of the "economic man" (Markley and Harman 1982: 47–8). Faith in progress toward material wellbeing and abundance was fueled by the power of science which allowed humans to control their destiny and the natural world. Modernism's turn to materialism was entrenched in the first half of the twentieth century as science provided ever more reductionistic accounts of nature (Cantor 1988: 6). Scientific breakthroughs in physics and biology fed this sense of advancement which continued to unfold with biotechnology, computerization, and information systems (Cantor 1988: 335).

Along with technological developments came more effective weapons to dominate other nations and groups. However, the use of poison gas and atomic bombs drew into question the blithe assumption of human progress and dignity (Childs 2017: 20). Critics deeply distrusted the optimistic teleology that promised a better life, and so emerged the backlash of Postmodernism which challenged the objectivity and value of science. Possessing complex and difficult knowledge, accessible only to experts, was framed as elitist—a quality that had once been a virtue became a vice in the latter half of the twentieth century.

***

Although early twentieth-century Modernism sought universal laws and morals, the First World War rocked the foundations of an all-encompassing philosophical framework. Turning away from theoretical endeavors, *pragmatism* emerged with its endorsement of radical empiricism. Truth could be known only by testing an idea in the world and objects could be understood in terms of their function or utility. The rigorous philosophical pragmatism of William James, John Dewey, and others (e.g., fallibilism rather than skepticism and perspectivism rather than relativism) eroded into a kind of short-sighted, "whatever works" approach.

Pragmatism complemented scientific frameworks, as the former adopted the conceptual approach of evolutionary biology as the process through which new ideas developed such that only the most fit survived (i.e., those which provided the desired results). Along with this came the pursuit of efficiency and productivity which gave rise to massive systems of industrial manufacturing (e.g., multinational corporations) and global institutions (e.g., the United Nations). To support these sprawling endeavors, western culture developed increasingly complex forms of administration (e.g., multi-layered bureaucracies with professional managers) and surveillance (e.g., private and government monitoring of citizens) (Childs 2017: 16–17).

***

The twentieth century began at the end of the Victorian era, a time in which a rigid system of values provided the foundation for faith in historical progress and nationalistic loyalty (Cantor 1988: 6). Christian ethics placed humanity and reason at the center of the cosmos, and this served as the basis for a sense of individual duty and patriotism (Cantor 1988: 132; Childs 2017: 16–17). But the carnage of the First World War led to an existential crisis in western culture. It was no longer practical to believe in continuous progress toward universal morality. The battlefields of Europe made clear that rationality and decency were far from assured.

What emerged and has largely prevailed in a cultural context for most of the twentieth century was moral *relativism*, most often expressed in terms of ethics being whatever a society deems to be right or wrong. This view aligned with a kind of degenerate pragmatism and served to provide a Postmodernist refutation of imperialism. The cultural despair fed by the horrors of the Second World War manifest with Postmodernists denying fixed meaning, literary canons, traditional values, or absolute truths (Cantor 1988: 35–9, 345).

***

The Industrial Revolution spawned social chaos which Romanticism critiqued and Victorianism sought to organize into culturally stable structures, thus forming the foundation for *corporatism*. The early twentieth century saw a conflicted relationship with technology, but the overall sense was one of optimism (Childs 2017: 19). However, when the instruments and carnage of war undermined the prevailing hopefulness, pragmatism and relativism supported an alternative view. Corporate structures offered the promise of a neo-Victorian, macrocosmic organization of human industry providing a renewed sense of social order and material efficiency.

With the horrors of war having devalued human life, individuals were seen as the means to the ends of military and economic victory. Given the importance of industrial production in defeating an enemy, not only soldiers but workers were part of the social machinery. Modernism was manifest through a Structuralist approach which moved the focus of reality and value from the individual to the system (Cantor 1988: 345). And with the rise of capitalism, technology rapidly transformed the infrastructure of society with ever newer fuels, medicines, cars, telecommunications, and synthetic materials (Childs 2017: 16–20). People became instruments of production, consumption, and ultimately corporate flourishing.

ENTOMOLOGICAL WARFARE

Entomology is an enormously rich field of basic and applied science which influences virtually every aspect of human culture. Combine this with the threads of Modernism and the result is a staggeringly complex history. As such, to provide analytical depth, it is necessary to narrow the scope of inquiry. As noted above, many of the elements of Modernism shifted in response to the realities of combat. War has the capacity to focus human perceptions and values and "test" the social limits of those engaged in the fight—and those who participate through socioeconomic connections.

The combination of entomology, health, and warfare in the twentieth century yields an intriguing context for analysis—the history of entomological warfare.[1] Insects had been used for centuries as weapons to inflict direct harm (e.g., via biting or stinging) and destroy crops (e.g., introduction of agricultural pests). However, with the discovery at the opening of the twentieth century that insects could transmit human diseases, the weaponization of these disease vectors became a potent element of modern warfare, at least in some conflicts.

Although studies of Modernism tend to focus on two "scientific revolutions" in the twentieth century—physics which culminated with the atomic bomb dropped on Hiroshima and biology which culminated with the discovery of DNA and the manipulation of the genetic code (Cantor 1988: 93)—in terms of military technology, entomological weapons were arguably more devastating than any other effort to control the forces of the natural world.

THE FIRST WORLD WAR

In the US Civil War and earlier conflicts, savvy strategists forced the enemy into swampy habitats where diseases were prevalent, but nobody understood that insects were responsible for the devastating outbreaks (e.g., "malaria" was derived from the Italian *mala aria*, meaning "bad air"). With the discovery in 1900 that mosquitoes transmit the virus that causes yellow fever, militaries were poised to conscript insects as allies in war. The scientific understanding of insect-borne disease allowed military commanders to knowingly engage in "passive" entomological warfare (Lockwood 2009: 65–76). In a passive strategy, the military leader takes advantage of naturally occurring insect-borne diseases by selectively protecting his own troops. The vectors are not added to the battlefield but they flourish in the mayhem and filth of war.

The scientific link between typhus and lice was made in 1909 and earned Charles Nicolle the Nobel Prize in Medicine. And so, in the First World War—a conflict in which the ideology of nationalism was emerging as a dominant social structure (Kohn 1971: 29)—the key to sustaining a healthy fighting

FIGURE 2.1 Dorsal view of a male body louse; the dark mass inside the abdomen is a previously ingested blood meal. The insect's legs and flattened body are well adapted to avoid being dislodged from their hosts, so during the First World War entomologists advised the troops to keep their hair short and faces shaved, change clothing often, keep infested uniforms away from their quarters and wear silk underclothes—to deny the insects a reliable foothold. Photograph by James Gathany, courtesy of Centers for Disease Control and Prevention.

force was hygiene. Because the major insect-borne disease depleting troops was typhus, controlling louse infestations in the trenches and quarantining sick soldiers were vital to success on the Western front (Lockwood 2009: 77–86).

The US Army Medical Corps was established in the First World War and provided the technical capacity to control illness (Heard 1950: 212). This operation was enormously successful and expanded in subsequent conflicts. While treating insect-borne disease was becoming an essential feature of modern warfare, the most efficient and effective tactic was prevention. In this context, the most important unit in the First World War was the Quartermaster Corps, which provided laundries for delousing clothing and stations for delousing soldiers.

The demonstrable efficacy of applied science in humanity's battles against nature and one another was driven home on the Eastern front where hygienic and medical interventions were lacking. Prisoners of war were kept in abhorrent conditions and there were only about four hundred physicians in all of Serbia. Filthy camps and ravaged villages fed a typhus outbreak that raged across the region for years, eventually killing as many as ten million people by 1921.

With respect to the threads of Modernism, these radically different experiences with insect-borne disease on the Western and Eastern fronts wove together the elements of ideology and scientism. The former served as the foundation for the nationalism that provided the cultural framework for the conflict. The latter demonstrated the capacity of scientific understanding to provide an enormous advantage to those who possessed expertise and could act on this knowledge. Complementing the power of medical entomology in battling typhus, the war also saw the development of poison gases, flame throwers, aerial combat, depth charges, machine guns, tracer bullets, and tracked tanks. Although the war called into question the Modernist ideals of universal morals and social progress, there was no doubt that technological advances were allowing the control of the natural world and shaping western culture in powerful ways.

THE SECOND WORLD WAR

The Second World War manifested the ideologies of nationalism, racism, and imperialism in unambiguous ways. With mechanized warfare, the scale of conflict expanded and there arose a need to subvert individuality and autonomy through appeals to existential danger. A collective belief in national superiority allowed governments to amass a fighting force and unify a homeland that would support the collective violence (Heard 1950: 73; Shannon 1974: 244). The invention and subsequent development of ever more sophisticated armaments (e.g., jet engines, radar, sonar, aircraft carriers, tanks, assault rifles, rockets, and atomic bombs) instantiated the idea that technological progress and specialized expertise were vital to assuring military victory and ultimately cultural survival (Markley and Harman 1982: 74).

***

Western countries treated Asian people with racial contempt, portraying them as slant-eyed, immoral schemers, despite the decency shown by the Japanese in the Russo-Japanese War (Lauren 1988: 78–80). In response, Japanese propaganda emphasized Asian unity against white domination, despite both Japan and China seeking imperial domination of the continent (Schroeder

1969: 317; Furedi 1998: 189). Japan's anti-colonial rhetoric was widely embraced, and the slogan of "Asia for the Asiatics" resonated with many individuals, although the Japanese oppressed their neighboring people (Furedi 1998: 166). When the Americans highlighted this inconsistency, the Japanese pointed to the racial relations in the US (Furedi 1998: 166). Whether Japan was racist in the conventional sense is disputable, but their sense of cultural superiority was exemplified by their treatment of the Korean people (Shin 2006; Rees 2017). The colonial government pursued a policy of assimilation, ordering all instruction in Korean schools to be in Japanese and pressuring Koreans to adopt Japanese names (Shin 2006).

Japanese imperialism flourished in the fertile ground of twentieth-century Modernism, and a nationalistic ideology paved the way for an entomological warfare project of unprecedented scale. The central player in this venture was Ishii Shiro who tapped into a cultural system of entitlement and rose through the military ranks as a physician and researcher. After he discovered that Japanese B encephalitis was transmitted by mosquitoes, Ishii realized the potential power of insect-borne disease. When Ishii considered the Geneva Convention's prohibitions on chemical and biological warfare—a kind of neo-Victorian effort to generate universal moral constraints—he inferred that these weapons must have tremendous potential or they would not have been banned. Although the Japanese military was not interested in biological weapons at the time, Ishii toured the world to gain information about these armaments. When he returned in 1930, Japan had changed dramatically, adopting an ultra-nationalist perspective that justified aggressive expansionism. And the potency of industrial science was increasingly appreciated (Lockwood 2009: 87–94).

Ishii convinced his superiors to support his research under the guise of protecting Japanese troops from illness, although his ultimate goal was to initiate, not prevent, epidemics. His work began with laboratory animals, but it was soon clear that further advances required experiments with human subjects. Radical nationalism and cultural chauvinism combined with moral relativism to pave the way for depraved science. Having left the League of Nations, Japan felt largely unconstrained by world opinion in the absence of universal values (Rees 2017). Ishii was free to do whatever was practically necessary to achieve victory over a culturally inferior enemy.

In 1932, Ishii moved his operation out of Tokyo, where the potential for an accidental release of a pathogen was unacceptable. He settled in Manchuria (which was then the puppet state of Manchukuo), where the military police could arrest virtually anyone and provide Ishii's research staff with human guinea pigs. He set up shop in Harbin, but soon realized that secrecy required a more isolated site. The Japanese High Command provided him with a new facility at Beiyinhe. The Zhong Ma Prison Camp could hold a thousand inmates for experimental purposes. In 1936, escaped prisoners revealed the nature of the secret project which made it necessary to either abandon this line of weapons development or move the operation to a new location. When it was discovered that the Russians were also working on biological warfare, the Japanese decided to expand their own program.

Ishii's Unit 731 occupied the newly constructed facility at Pingfan in 1938 (Lockwood 2009: 95–107). A fifteen-foot wall topped with high-voltage lines and barbed wire surrounded two square miles of land with more than 150 structures including: administrative offices, laboratories, a power station, a brothel, recreational facilities, housing for three thousand scientists, dormitories for technicians and soldiers, a prison for the inmates, and a crematorium. Pingfan instantiated the Modernist elements of corporate pragmatism—a carefully planned, systematically organized, highly structured operation with layers of bureaucracy. The facility was designed for efficiency and productivity with the expectation that Unit 731 would provide the Japanese military with potent weapons.

The official cover story was that the walled compound held a lumber mill, and from this ridiculous attempt at deception came the term used for the human subjects. The scientists referred to them as *maruta*, meaning "logs" which exemplified the cultural relativism that allowed the researchers to vivisect their fellow humans. People were reduced to instruments for scientific investigation and technological advancement—they were the means to the ends of nationalistic flourishing.

The first battlefield test of Unit 731's arsenal took place at Nomonhan against the Russians in 1939. The Japanese dumped typhoid and salmonella bacteria in the river and exploded plague-laden shells over the enemy—and nobody was sickened other than some of the Japanese who mishandled jugs of microbes. Ishii presented the campaign as a success to the clueless leadership, but he knew that his program would need a genuine accomplishment in the near future. In starkly pragmatic terms, results mattered—and "results" meant dead enemies.

The researchers understood that the problem was that microbes are unable to seek out hosts and the exposed pathogens died quickly in the environment. Vectors would solve these problems and with enough technological brute force the scientists could mass produce infected carriers. Although various pathogens had been tested by Unit 731, the scientists began to focus on bubonic plague and the fleas that carry the bacterial pathogen.

The Japanese began by rearing fleas on their prisoners and then dropping flea-infested rats onto enemy targets. However, this system lacked industrial efficiency. The scientists found that fleas could survive being encased in a porcelain bomb shell that exploded above the ground and rained the insects onto the target, eliminating the need for rodents to be involved in an attack.

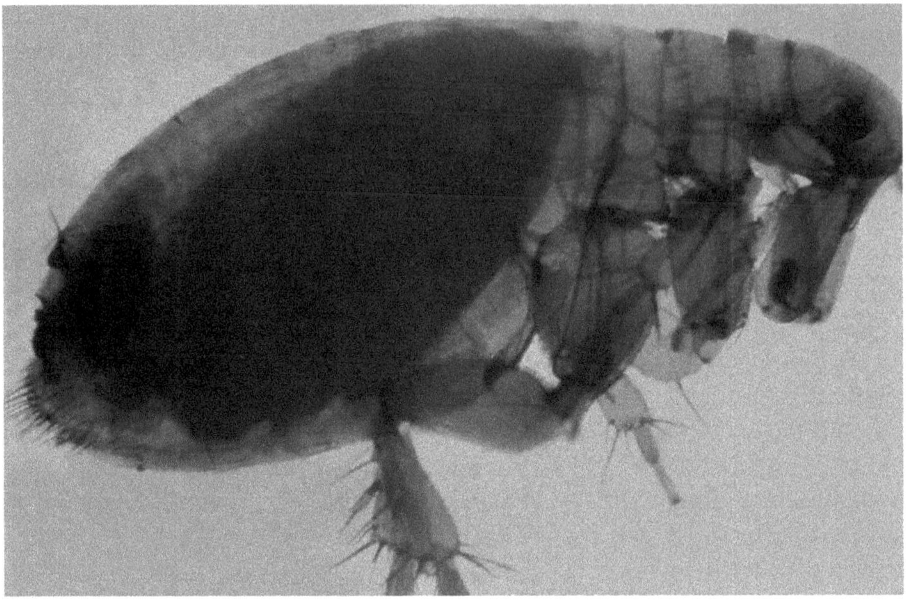

FIGURE 2.2 The oriental rat flea is one of the insect vectors of bubonic plague, and it was the key to Unit 731's entomological weapons program. This photograph shows a mass of bacteria in the insect's digestive system (the dark mass just in front of its blood-filled gut). This blockage prevents the hungry flea from ingesting a meal and forces the insect to regurgitate infected blood into the host, thereby spreading the pathogen in its futile efforts to feed. Photograph courtesy of Centers for Disease Control and Prevention.

Soon thereafter the scientists found that the insects could be directly sprayed from airplanes. Unit 731 devised methods for the mass production of plague-infected fleas. A factory with 4,500 incubators holding rodents infected with the bacteria served as the blood and pathogen source for the insects. With this innovation, Ishii's entomological warfare program was producing five hundred million infected fleas a year.

This research and development venture, like that of a major corporation devoted to manufacturing a product, required considerable organizational skill as well as ethical rationalization. With between six hundred and two thousand *maruta* used for experiments each year, moral sensibilities eroded among the researchers with the aid of lucrative salaries and nationalistic values. Allied POWs were included in tests, and the ideology of Japanese superiority made it possible to use these humans as guinea pigs. If morals are culturally relative and the Emperor condoned the goals of Unit 731, then it followed that the research was the right thing to do.

The work conducted by Ishii's scientists was communicated to the highest levels of the government and military, along with being published in scientific journals. The human subjects were relabeled as "Machurian monkeys"—a non-existent species—and results were published in the *Japan Journal of Pathology* showing infected animals with fevers of 104.4^{0}F (a higher body temperature than any monkey attains). The scientific community was aware of the nature of the experiments being conducted at Pingfan, but the darkest elements of Modernism conspired to keep the research an open secret.

Unit 731 had satellite facilities across eastern China, where research was conducted on weaponizing typhus-infected lice and Songo-infected ticks. This expansion would not have been possible without militarily significant results. Given political uncertainties, the Japanese were initially hesitant to operationalize entomological weapons, but their use of poison gas against the Chinese in 1937 produced no serious response from the League of Nations. So with a Modernist combination of imperial ideology, moral relativism, and crass pragmatism, Unit 731 began its campaign against Chinese targets. In 1940, plague-infected fleas were spread over three cities. The most successful mission took place at Quzhou, where thousands died in the outbreak. Over the next two years, Unit 731 coordinated more than a dozen attacks inflicting over a hundred thousand casualties (Lockwood 2009: 108–16).

With these successes, Unit 731 expanded its program to include a new disease system. To punish the Chinese for protecting American pilots who had bombed Japanese cities and then landed in the region of Zhejian, the Japanese released plague-infected fleas and dumped cholera bacteria into the water supply. The latter approach produced more Japanese casualties from handling the pathogens than the Chinese suffered from the contaminated water. The problem was that the bacterial populations were rapidly diluted. Unit 731 required a means of triggering an initial wave of infection which could then spread through fecal contamination. What they needed was a vector.

The Japanese devised the binary Yagi bomb in which a slurry of cholera bacteria drenched a compartment of house flies upon impact as the casing burst open. The cholera-coated flies then spread and contaminated households. To increase the efficacy of this approach, the Japanese developed a staged assault with the first wave of bombers dropping conventional ordinance to decimate the infrastructure of Baoshan—a key city in Yunnan Province through which an Allied supply line was providing arms to Chiang Kai-Shek's Nationalist forces. After the explosive and incendiary bombs came the Yagi bombs. Once the disease had been established in the urban populace, the Japanese dropped conventional ordinance to drive the people out of the city and into the countryside and thereby spread cholera. The outbreak swept across an area equal to that of Pennsylvania, killing two hundred thousand people, contaminating the Allies' supply line, and creating a diseased no-man's land. In 1943, the same "decimate-and-contaminate" strategy was used in the northern province of Shandong. The twist was that Japanese troops were in the area, so before the attack the soldiers were vaccinated against cholera. The ensuing epidemic caused 210,000 deaths.

Despite Japan's entomological weapons, by 1944 it was evident that the war was turning (Lockwood 2009: 117–27). With the Russians preparing to invade Japanese-held territories in China, Unit 731 planned for the production of five billion fleas a year. Infected blood became a limiting factor, so the *maruta* were kept in kennels, injected with plague bacteria, allowed to sicken, and then exsanguinated to feed fleas. While the Japanese devalued the lives of their Chinese captives, the Americans discounted the danger of the Japanese developing technologically sophisticated weapons because of racist assumptions as to the intellectual capacity of Asian people (Furedi 1998: 168).

This ideological arrogance nearly cost the Allies gravely. Several planned attacks on US targets in the Pacific failed for logistical reasons, and a major assault on San Diego with plague-infected fleas was aborted when a Japanese general saw that biological warfare could spiral into an endless, global battle of disease. A kind of neo-Victorian shame with what this would mean for Japanese culture prevented him from ordering the attack.

While technology was producing entomological weapons for Japan, science was providing nuclear weapons for the US. In the waning days of the war, the Japanese slaughtered their remaining human subjects and released infected rats into the countryside. General Ishii was captured by the Americans, who agreed not to try him as a war criminal in exchange for his technical knowledge. The Americans had underestimated the Japanese and fallen behind in terms of biological weaponry. The Russians captured several key scientists from Unit 731 and convicted them of war crimes, but then released them in exchange for their information. The pragmatic trade of technology for justice by the Americans and Russians was Modernism at its best—or its worst.

***

Germany's entomological warfare program was much less advanced than Japan's, in large part due to Adolph Hitler's aversion to unconventional weapons after having been gassed in the First World War (Lockwood 2009: 128–38). Nonetheless, the Nazis pursued the development of insect-borne diseases and made some inroads with regard to weaponizing typhus-infested lice using concentration camp prisoners as experimental animals. As in Japan, Modernism's elements provided a combination of ideological nationalism/racism along with cultural relativism to sweep aside moral compunctions (Lauren 1988: 118–19). And these factors worked in concert with the intense scientism to feed the technological innovations.

Perhaps the most vivid example of Modernism's dark capacities emerged as the Germans sought to find a way to protect their crops from what they mistakenly took to be an impending attack with Colorado Potato beetles. In fact, the German scientists were far ahead of the Allies in weaponizing insects to destroy the food supply of an enemy. The research effort to create new insecticides led to the discovery of the organophosphate compounds which turned out to be both effective chemicals for pest management (e.g., today's malathion is based on this chemistry) and the basis for nerve gases (e.g., sarin, which was used in the 2013 attacks on Ghouta by Syrian forces).

***

The elements of Modernism drove Germany and the Soviet Union, two very different nations, toward convergent technologies (Lockwood 2009: 128–38). In an effort eerily similar to that of the Nazis, Soviet scientists pursued the development of entomological warfare through the weaponization of lice to spread typhus among enemy troops. Some work was also conducted with plague and cholera (as in Japan's Unit 731), along with anthrax (as in Britain's biological warfare program) and tularemia (commonly transmitted by deer flies, horse flies, and ticks). Meanwhile, the Soviets' search for insecticides yielded phosgene oxime or nettle gas—a potent chemical weapon.

The Germans accused the Soviets of having started an outbreak of tularemia outside of Stalingrad in 1942 and insinuated that the typhus outbreak which ravaged German forces on their retreat was also instigated by the Russians. These claims cannot be verified, but in 1941 the Soviets were conducting research with plague-infected fleas on political prisoners and Japanese POWs in Mongolia. When a diseased inmate escaped and triggered an outbreak in nearby villages, the Soviets conducted airstrikes on the afflicted settlements to prevent an epidemic.

***

The British were concerned about the potential that yellow fever (a mosquito-borne disease) would be introduced to India by the Axis. As a defensive maneuver, the British stockpiled vaccine for their troops—and then shifted their efforts to offensive capabilities. The focus was on anthrax and the British collaborated with the US where there was the industrial capacity for the mass production of anthrax bombs (Lockwood 2009: 139–58).

Frederick Banting, who shared the 1923 Nobel Prize in Medicine, was unable to secure government funding in Canada for biological weapons so he tapped into private corporations. Through connections to the leaders of the Canadian Pacific Railway, the T. Eaton Co., and Seagrams liquor, Banting raised more than $1 million to support a research program in a Modernist integration of corporate and technological prowess. The Canadians worked with American scientists to test the potential for delivery of pathogens with flies and mosquitoes on Horn Island off the coast of Mississippi. The Canadian's work continued after the war, evolving into a device similar to Japan's Yagi bomb. Research in Canada culminated in the development of house-fly baits laced with pathogenic bacteria, which could be dropped into enemy zones where the local insects would distribute the microbes (Lockwood 2009: 139–58).

***

Like the other countries fighting in the Second World War, the US had developed an ideology of nationalism which the Americans founded on their country being unique and exceptional with regard to democratic pluralism (Shannon 1974: 244; Cantor 1988: 92–3). Until the 1940s, the US denied the viability of biological warfare due to the environmental vulnerability of the microbes and the risk of infecting one's own troops. When military intelligence revealed that the Japanese were waging entomological attacks in China, the Americans accelerated their research and development program (Lockwood 2009: 139–58).

The biological warfare effort spawned the Entomological Warfare Division at Camp Detrick in Maryland. Researchers pursued methods for the mass production of fleas, mosquitoes, and lice. The scientists parlayed the arrival of Japanese balloon bombs to the west coast of the US into an opportunity to boost funding. They speculated that the enemy could deliver Japanese B encephalitis in a balloon payload—a pointless tactic in the absence of associated vectors. However, one of the "advantages" of technical expertise in Modernist culture is that politicians and other non-experts can be easily duped with pseudo-scientific claims.

The most impressive control of nature was realized in the American programs to protect their troops and allies from insect-borne disease. Fearing that a typhus outbreak among the Italian populace would spread to Allied troops, the US Typhus Commission was formed in 1942 and the weapon of choice was DDT. The commission organized a massive delousing program, and within a few months the US army had dusted 3.3 million people with 127 tons of DDT powder and thereby avoided an epidemic that would have bogged down the invasion.

In the Pacific, General Douglas MacArthur declared, "This will be a long war, if for every division I have fighting the enemy, I must count on a second division in the hospital with malaria and a third division convalescing from this debilitating disease" (Lockwood 2009: 154). The Americans developed specialized equipment that used a propellant to force the oil-based insecticide through a fine nozzle and loaded DDT into B-25s and C-47s equipped with 625-gallon tanks. Before the program began, the malaria rate among Allied forces was 208 cases per thousand troops; this rate plummeted to just five cases per thousand troops within two years.

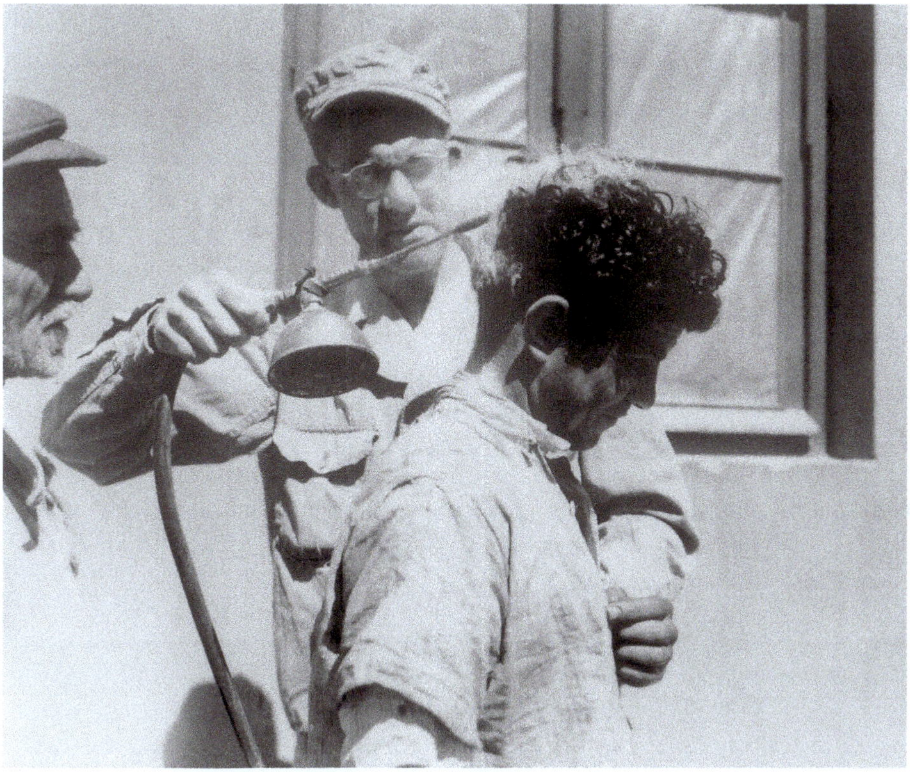

FIGURE 2.3 A US soldier uses a power duster to apply DDT powder in the battle against louse-borne typhus, most likely in Italy during the summer of 1943. The Typhus Commission dusted more than three million people, saving thousands of civilian lives and assuring that a healthy Allied force could continue the push into northern Italy. Thanks to DDT spraying in the Pacific theater, the malaria rate among Allied forces dropped by 98 percent from 1943 to 1945. Photograph courtesy of Centers for Disease Control and Prevention.

THE COLD WAR

The entomological weapons developed and deployed during the Cold War entrenched the elements of Modernism (Alperovitz 1969: 390–8; Schlesinger 1969; Cantor 1988: 35–9). The global conflict was among competing economic, imperialistic ideologies; communism and capitalism vied for world domination. Science promised to provide the technological means to win the arms race—whether the weapons were missiles or vectors. In pragmatic terms, the ends of victory justified whatever means were necessary, including the spread of disease into the enemy's civilian population. Although the United Nations made an admirable attempt to provide countries with a common set of ethical constraints, a kind of moral relativism in which societies placed different values on freedom, equality, and individuals made such unification impossible.

***

The early work with biological warfare agents at Camp Detrick was not highly promising. The initial goal was to impose the maximum technological control over nature. The reductionistic model of science dictated that single pathogens should be mass produced in highly controlled, industrial processes and then released as an aerosol into the environment, much like a poison gas would be distributed. In essence, microbes were treated as chemicals. The scientists avoided the messiness of vectors because insects were difficult to produce, infect, handle, store, and disperse. However, testing soon drew the US scientists to the same conclusion as their Japanese counterparts—without a vector, the microbes did not survive, disperse, or infect in ways that were militarily viable (Lockwood 2009: 139–58).

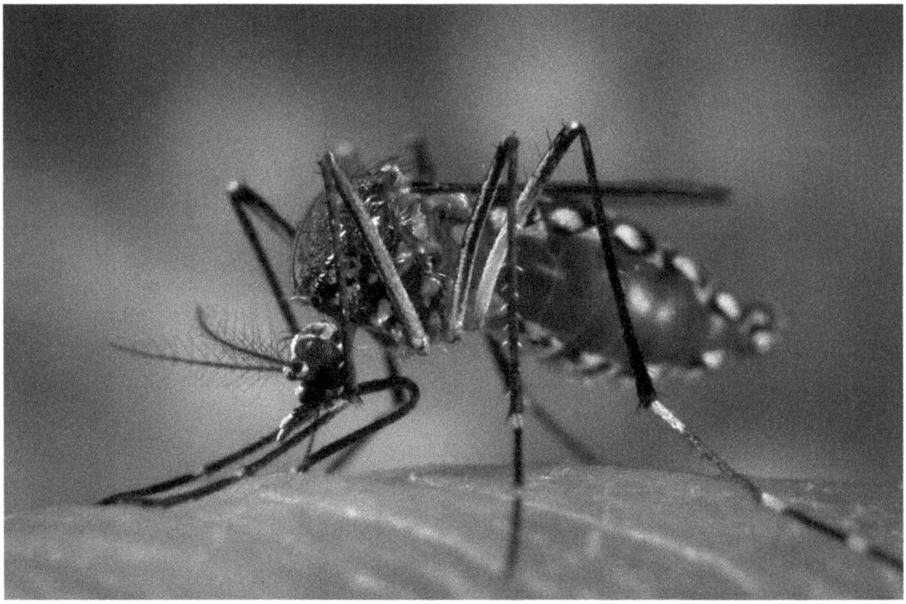

FIGURE 2.4 A female of *Aedes aegypti*, the mosquito that transmits the virus causing yellow fever. This vector/disease system became the mainstay of American's entomological warfare program in the 1950s when scientists at Fort Detrick discovered that the insect could uptake the virus directly from the larval medium. Plans were drawn up in 1960 for a facility to produce as many as a hundred million infected vectors per week at Pine Bluff Arsenal in Arkansas. Photograph courtesy of James Gathany/Centers for Disease Control and Prevention.

The research at Camp Detrick initially focused on plague-infected fleas based on knowledge gleaned from the Japanese. In 1954, field testing with uninfected fleas was conducted at Dugway Proving Grounds in Utah under the code name of Operation Big Itch. The insects were released from E-14 and E-23 munitions dropped from one thousand to two thousand feet and the fleas readily infested the caged, test animals. Plans were developed for a facility to produce fifty million fleas/week, but mass culturing the plague bacteria became a serious obstacle.

Detrick was stocked with mosquitoes carrying the pathogens that cause yellow fever, malaria, and dengue; flies with dysentery, cholera, and anthrax; and ticks with tularemia, relapsing fever, and Colorado fever. The focus of the program shifted to yellow fever when it was discovered that larvae of *Aedes aegypti* could be directly infected by adding the virus to their watery medium, thereby allowing mass production of infected insects. A series of field releases with uninfected mosquitoes was conducted. For example, Operation Big Buzz involved the release of more than a million mosquitoes dispersed from E-14 munitions over rural Georgia. The insects survived the release and located hosts (volunteer humans) a half-mile from the drop site. With this success, the Americans became concerned with protecting the country from a Soviet entomological attack, particularly in poor, southern states where there was limited medical capacity and inadequate equipment for spraying insecticides.

Until 1956, the US had an ambiguous position with regard to using biological weapons. However in that year, the Soviets announced their willingness to use biological and chemical warfare in future conflicts. In response, the US National Security Council declared: "To the extent that military effectiveness of the armed forces will be enhanced by their use, the US will be prepared to use chemical and bacteriological weapons in general war" (Lockwood 2009: 197). The moral position of ethical relativism combined with stark pragmatism to justify retaliation-in-kind and to leave open the possibility of first use.

Camp Detrick was elevated to Fort Detrick and entomological experiments were intensified. Releases of uninfected mosquitoes over residential areas of Georgia made the public into unwitting participants in a grand ideological project to secure American victory. A new dispersal system was developed and tested with two hundred thousand mosquitoes over a Florida bombing range. The XM28 "bagged-agent dispenser" worked well and was subsequently used in experiments over populated areas with mosquitoes spreading up to two miles before biting people.

The Bellweather tests in 1959 investigated the environmental factors under which releases of various mosquito strains would be most effective. These data contributed to models intended to achieve scientific predictability. An "Entomological Warfare Target Analysis" by the US Army Chemical Corps in 1960 revealed which cities in China, the USSR, and Soviet-allied countries would be optimal locations for an attack with yellow fever mosquitoes.

By 1960, the entomologists at Fort Detrick could produce five hundred thousand infected mosquitoes per month. Plans were drawn up for increasing production by 250-fold in the world's largest insect-rearing facility at the Pine Bluff Arsenal in Arkansas. The annual operating budget was to be $5 million with nearly two thousand workers staffing a site that would cover more than two square miles and include factory buildings with 45,000 gallon stainless steel vats to mass produce infected insects with industrial efficiency. The plans for an enormous mosquito-production facility in preparation for waging entomological warfare were never realized but not because of moral compunctions, technological problems, or practical concerns. Rather, in the 1960s engineers had come up with munitions that would release microbial aerosols of optimally-sized particles which promised greater control and reliability.

***

The structuralist tendency toward multinationalism (e.g., the United Nations) came into tension with nationalism and relativism—and this conflict became evident with accusations that the US had waged entomological warfare in Korea (Kohn 1971: 132; Cantor 1988: 345). The war began on June 25, 1950 and rapidly became a stalemate. The US had for some time been investigating mosquito-borne diseases at its 406 Medical General Laboratory in Yokohama, with assistance from scientists associated with Japan's Unit 731. This pragmatic affiliation and connections to Camp Detrick led to rumblings that offensive capabilities were being developed (Lockwood 2009: 159–70).

In February 1952, the North Koreans issued an official accusation to the UN Secretariat that the US had released "large quantities of bacteria-carrying insects" and the Chinese added their own charges that hundreds of aircraft had dropped contaminated insects (Lockwood 2009: 171–81). American officials flatly denied the accusations. That spring, the International Association of Democratic Lawyers (with a leftist ideology) took up the case and issued a report filled with biological absurdities and photographs of misidentified insects (e.g., "poisonous insects" that were stoneflies).

The US asked the UN to conduct a full inquiry and the Red Cross was chosen to investigate the charges. However, North Korea and China denied inspectors access to sites and information. Meanwhile, the Soviets pointed out that the US was the only member of the UN Security Council that failed to ratify the Geneva Protocol prohibiting the use of biological weapons (ratification finally came in 1975). And so ideology trumped science, the Red Cross abandoned its investigation, and the Soviets vetoed a UN resolution exonerating the US.

China then turned to another ideological ally, the World Peace Council, to pursue the charges (Lockwood 2009: 182–92). The Council issued a report documenting the use of fourteen different arthropods infected with eight pathogens on thirty-three separate occasions including: *Aedes* mosquitoes (unknown pathogen), anthomyiid flies (anthrax, cholera, dysentery, paratyphoid, and typhoid), crane flies (neurotropic virus), culicine mosquitoes (unknown pathogen), green bottle flies (unknown pathogen), house flies (anthrax), midges (typhoid), stable flies (typhoid), sun flies (paratyphoid), human fleas (plague), ptinid beetles (anthrax), grasshoppers/locusts (unknown pathogen), springtails (dysentery), stoneflies (unknown pathogen), and wolf and lycosid spiders (anthrax and fowl cholera).

The report included first-hand accounts from villagers of insects raining from the sky and objects dispersing insects. There was also a litany of ecological anomalies (e.g., springtails on concrete), temporal anomalies (e.g., locusts on snow), extraordinary densities (e.g., mounds of house flies and crickets), strange associations (e.g., springtails and fleas together), peculiar occurrences (e.g., flea-infested voles being parachuted from planes), and insect-microbe anomalies (e.g., ptinid beetles with anthrax) all of which pointed to entomological weapons. Stories of finding plague-infected fleas were reminiscent of well-documented attacks by Unit 731 during the Second World War. There were also rambling handwritten confessions by captured US pilots.

The US response was a point-by-point rebuttal explaining away most of the "evidence" as being natural events or absurd biology (e.g., no rational military would use ptinid beetles to spread anthrax). However, the refutation overreached in claiming that the US had never investigated the use of arthropods for biological warfare, given that the Entomological Warfare Division had been operating at Detrick for more than a decade. In addition, the US attorney general coerced the recantations of the American POWs by threatening charges of treason, which would have been unnecessary had there been no truth in their statements. Politics aside, science trumped ideology in the minds of most observers. Nonetheless, the debate continued until 1998 when documents showed that North Korea, China, and Russia conspired to falsely accuse the US of having waged entomological warfare—and most experts now believe that such weapons were not used in large-scale attacks.

***

Entomological warfare took a couple of strange twists during the war in Vietnam (Lockwood 2009: 203–9). On the American side, herbicides (Agent Orange) were sprayed in Operation Ranch Hand to denude two thousand square miles of forest in 1961. This program to deprive the Viet Cong of cover triggered an unexpected outbreak of scrub typhus among the enemy. When grassland and secondary forests regrew, these habitats fostered rats which carried mites which transmitted the pathogenic bacteria. The Americans did not plan for this result, but they did plan to protect their troops from bubonic plague knowing that the enemy would be exposed to the disease through the filth of war. Through a hygiene and vaccination campaign, only a dozen American soldiers contracted plague during the war while this flea-borne disease afflicted twenty-two of the twenty-nine provinces north of Saigon. For their part, the Viet Cong conscripted stinging insects. Colonies of the giant honey bee were relocated along trails as booby traps and boxes of scorpions were placed in the tunnels of Cu Chi to ambush unwary Americans.

***

The 1980s and 1990s saw the Cubans making continual accusations regarding the use of entomological weapons by the US as ideology and technology clashed in the Caribbean (Lockwood 2009: 210–20). An outbreak of mosquito-borne dengue in 1981 was blamed on the Americans based on circumstantial evidence along with a wild claim that the US had used its scientific prowess to alter the weather and thereby foster the production of mosquitoes.

FIGURE 2.5 During the Cuban Missile Crisis, the Americans considered unleashing planthoppers against Castro's crops. Entomologists at Fort Detrick had been working with the US Department of Agriculture to weaponize planthoppers and leafhoppers as vectors of plant diseases to destroy sugarcane and rice. This photo shows the black-faced leafhopper, which transmits diseases of corn. Photograph courtesy of USDA/Stephen Ausmus, https://www.ars.usda.gov/oc/images/photos/may05/d075-28/.

At the time, the US had conducted "An Evaluation of Entomological Warfare as a Potential Danger to the US and European NATO Nations" but the study focused on yellow fever, not dengue. Modernism's elements of scientific analysis, technological control, and corporate efficiency came together to reveal a cost of $0.03 per death—an impressively exacting calculation.

The Cubans also accused the Americans of releasing panicle rice mites, brown citrus aphids, and mites afflicting honey bees. While these claims are dubious, the US government did consider releasing insect vectors of plant pathogens during the Cuban Missile Crisis. Meanwhile, the Soviets asserted that CIA operatives were posing as malaria-eradication workers in Pakistan while spreading the mosquito-borne diseases of dengue and yellow fever.

The most serious charges were brought by the Cubans under Article V of the Biological and Toxin Weapons Convention which triggered a formal, international investigation (Lockwood 2009: 221–32). The US was accused of releasing insects over Cuba. An American plane had flown across the island and a thrips outbreak had decimated potato fields, but pests arrive on islands through a variety of natural means. In the end, the investigation yielded no definitive conclusion.

CONTEMPORARY AND FUTURE CONFLICTS

In today's world, Modernism provides us-them discord arising from cultural fragmentation, despite economic tendencies toward globalization (Cantor 1988: 35–9). Woven into this situation is a continuing belief that technological progress—increasingly dependent on corporate structures—will provide solutions. However, it should be noted that there is also Postmodernist disenchantment with technology given the misapplication of genetic, nuclear, and chemical products (Markley and Harman 1982: 74). Non-industrial nations (which are often conflated with racial ideologies of inferiority) exhibit ambivalence to western science given its role in colonial oppression (Heard 1950: 26; Kranzberg 1971). As such, the use of insects as weapons currently manifests in both simple and high-tech formulations.

***

Bioterrorism using insects is a serious matter. Although some of the actual and anticipated attacks use pests of agriculture, the greater concern pertains to insect-borne diseases (Lockwood 2009: 256–73). In a world with vast economic differences across borders, there has developed a profound asymmetry with regard to the capacity to harm one's enemy. With industrial countries having such enormous advantages in terms of technological sophistication, rogue nations and non-state actors understand that very simple tactics can offset this scientific imbalance. Just as box cutters were sufficient to hijack an airplane and cause tremendous damage during the 9/11 attacks, pathogen-laden vectors have the potential to function as "biological box cutters."

The criteria for a disease being suitable as a weapon include its being infective in small doses, causing acute and severe illness, retaining potency during production, storage and handling, and surviving in the environment after release. These features align well with many insect-borne diseases, as exemplified by the cases of Japan's plague-infected fleas and America's yellow fever-infected mosquitoes. An analysis by the Stockholm International Peace Research Institute (SIPRI) of likely pathogens to be used in biological warfare listed twenty-two candidates with half being arthropod-borne viruses. But one might wonder whether Modernism provides the antidote through a capacity to control nature. That is, won't the power of corporations to manufacture insecticides combined with the government structures to disperse these chemicals protect a populace from something as primitive as biting insects?

The answer to this question is empirically evident in the case of West Nile virus (WNV) which demonstrated the vulnerability of the US to a newly arrived, insect-borne disease. The illness was initially misdiagnosed as St. Louis encephalitis and a month passed before a correct identification was made. Despite a massive program of insecticide spraying in New York City based on the wrong disease (but the right vector, mosquitoes), WNV moved rapidly across the country. Having arrived in the summer of 1999, the virus spread to four neighboring states by the end of the year. In 2000, WNV was found in twelve states, in 2001 infections were reported in twenty-seven states, and by 2002 people in forty-four states were afflicted. In these first three years, the number of human cases increased from fewer than one hundred to nearly ten thousand with deaths rising from ten to 284.

The general assumption is that WNV arrived by accident, although the Centers for Disease Control and Prevention have not provided evidence to support this view. However, the Centers have denied that there was any intentional release. It is unlikely but not impossible that the pathogen arrived through Iraq's biological warfare program, given their intense interest in this disease, access to the virus, collaborations with the Russians, and occurrences of suspicious outbreaks in Israel and Saudi Arabia—two of Saddam Hussein's top enemies.

With regard to the next insect-borne disease capable of causing serious harm to industrialized nations, the best prospect may be Rift Valley fever (RVF). The virus is spread both by both aerosols from infected hosts and by mosquitoes. The first outbreak occurred in 1930 in Kenya and caused 10–70 percent mortality in cattle, along with livestock abortions. In 1950, a hundred thousand lambs and sheep were killed in an outbreak. And then in 1977, RVF crossed the Sahara into Egypt afflicting 25–50 percent of all

FIGURE 2.6 A pen of goats in Saudi Arabia during the 2000 outbreak of Rift Valley fever. The disease caused tens of thousands of sheep and goats to abort their fetuses, adding hunger and poverty to villages stricken by the illness. More than two hundred people contracted lethal cases of the disease, either via mosquito bites or through contact with the blood or other bodily fluids of infected animals. Photograph courtesy of Abigail Tumpey/Centers for Disease Control and Prevention.

sheep and cattle—and generating the first reports of human cases with neural and hemorrhagic forms. Some two hundred thousand people were sickened, leaving 1 percent blinded and 598 dead. Fort Detrick provided vaccine that stemmed the epidemic—demonstrating that this disease was of significant interest to US biological warfare experts. Subsequent outbreaks caused 200 deaths in West Africa (1987), 300 deaths in Kenya, Somalia, and Tanzania (1997), and 855 deaths in Saudi Arabia and Yemen, with tens of thousands of sheep and goats killed (2000).

Because RVF virus is carried in the eggs of mosquitoes, the disease would be easy to introduce to a new locale. As the director of the US National Center for Biodefense and a former researcher at Fort Detrick noted:

> ... simply go to an endemic area during an outbreak, collect floodwater *Aedes* [mosquitoes] from prime habitats, let them feed on a viremic [infected]

> animal, collect the eggs, put them on filter paper, let them dry, put them in your shirt pocket, come into the country, go to a suitable habitat, drop the filter paper into the water, and walk away.
>
> (Lockwood 2009: 267)

The expected mortality during an outbreak of RVF is ten times that of WNV, with up to 10 percent of survivors having permanent vision damage. The US has a number of mosquito species that would be suitable vectors. Knowing that these are potential carriers seems to be of little value in terms of stopping an epidemic. In 1985, the Asian tiger mosquito was found breeding in Houston—a newly arrived species in a limited geographic area with well-established ability to transmit disease. Nevertheless, within two years the insect had spread to seventeen states.

In the 1960s, scientists worked to breed cold-hardy mosquitoes which would extend the yellow fever weapon system into the northern reaches of the USSR. By the 1980s, SIPRI was predicting the development of more efficient and entirely new insect-microbe associations through the genetic engineering of vectors. And today, the former research leader of the USDA's Arthropod Borne Animal Diseases Research Laboratory maintains that, "there is no reason why a mosquito could not be genetically engineered to transmit, even perhaps produce, the HIV virus" (Lockwood 2009: 271). The converse is also possible and arguably easier—modifying the genetics of the microbe so that it fits into the physiological system of an insect carrier already abundant within the enemy's homeland. As yet another alternative, *New Scientist* published an article portraying a hypothetical scenario in which whiteflies were spreading through the American corn crop and transmitting a plant virus that was naturally associated with the insect, but had been genetically modified by terrorists to synthesize botulinum toxin (MacKenzie 1999). The result of such an attack was portrayed as hundreds of sick and dying people, billions of dollars lost in exports and control programs, and widespread panic in the populace.

***

It is difficult to guess what cultural framework will emerge after Modernism/Postmodernism, although at least one element of twentieth-century society seems likely to prevail for decades—the use of force and the application of power. Much of the last two hundred years has reflected two central objectives: the control of natural phenomena and the control of other people (Heard 1950: 207–11). With transnational corporations emerging as dominant social structures and democracies edging toward corporatocracies, there remains the desire to impose control—and insects can be expected to continue as weapons in our efforts to dominate both nature and one another.

The practicality of introducing insect vectors (or pests of agriculture) to the industrial world by those who have been aggrieved by the ideologies of capitalism, racism, and nationalism has shifted attention from offensive to defensive approaches in western countries. In this regard, both legalistic and technological approaches reflect the continuing influence of Modernism (Lockwood 2009: 298–310).

International laws, such as the Biological and Toxin Weapons Convention, forbid the use of living organisms as weapons (presumably including insects, although they are not named in most accords). Of course, non-state actors and rogue governments are not signatories to these agreements.

In terms of technology, the US and other countries have programs to detect and respond to an attack (or accidental introduction). However, the scales of international commerce and travel are such that inspectors are overwhelmed, with perhaps 0.01 percent of people and packages being examined. Most of the biodefense resources are directed toward avoiding an acute attack on human life. Technological sophistication with regard to surveillance and detection is the first line of defense, which may not align well with the very low-tech methods likely to be used by enemies (e.g., the introduction of RVF via mosquito eggs). Some of the most intriguing Modernist advances in detection are cyborgs, in which the phenomenally sensitive sensory systems of insects are integrated into machines.

While Modernism would seem to underwrite elements of how industrial nations are preparing for entomological terrorism, perhaps the most effective defense regarding the arrival of a new insect-borne disease (accidental or otherwise) is a strong public health infrastructure (e.g., vaccines, diagnostics, and educational programs). Although this pragmatic tactic has elements of Structuralism, it lacks radical progressivism. A public health clinic is not a high-tech hospital with specialists engaged in reductionistic practices. But perhaps public health captures elements of Postmodernism such as the importance of contextual, pluralistic, and culturally situated understandings—and the defiance of the grand narrative as to technology's heroic capacities. However, there may be an emerging grand narrative in the twenty-first century regarding the vulnerability of humans to global pandemics, as governments fail to provide the essential infrastructure to detect, prevent, or mitigate disease outbreaks—including those involving insects as naturally occurring or anthropogenically augmented vectors (Garrett 1995, 2001).

CONCLUSION

The development and use of entomological weapons to spread disease in the twentieth century can be viewed through the lens of Modernism and the responses to this dominant perspective. The elements of ideology, scientism, pragmatism, relativism, and corporatism interact to illuminate, if not entirely explain, how insects came to be instruments of war and terrorism. Of course, this is only one approach to understanding cultural history. One might equally well interpret the conscription of insects into human conflict in terms of the features of (de)colonization, cultural homogenization, economics, environmental degradation, medicine, communication, or even art.

On the other hand, entomological weapons are not the first objects that historians of the twentieth century would use to make sense of this tumultuous period. However, at least some of this oversight stems from there having been little study of how important insects were during these times. Few history books reveal that, as defining as the bombing of Hiroshima and Nagasaki were to the century, the Japanese killed more Chinese with biological weapons (580,000 deaths, with slightly more than three-fourths by entomological weapons) than the Americans killed Japanese with nuclear arms (225,000 victims) (Lockwood 2009: 127).

The philosopher George Santayana famously said that "Those who cannot remember the past are condemned to repeat it." What might be more worrisome when it comes to weaponizing insects or other organisms is the possibility that those who know history are empowered to recreate it—in even more devastating terms.

CHAPTER THREE

Insects and Food

DAVID KINKELA

> An infertile woman and an infertile field both receive a dose of chemicals, at a calculated risk of undesirable consequences, and are thus reduced to the status of productive machines. And for unwanted life—sperm, ova, embryos, weeds, insects, etc.—we have the same sort of remedies, for sale, of course, and characteristically popularized by advertisements that speak much of advantages but little of problems.
>
> (Berry [1977] 2015: 138)

In the twentieth century, food production in the United States, as Wendell Berry lamented, became enmeshed with a near zealous quest to industrialize nature. In his influential book *The Unsettling of America*, Berry criticized the rapid transformation of agricultural landscapes, which became more factory than farm with each passing year. According to Berry, America's embrace of industrial capitalism was the key to understanding this change. The tenants of profits, efficiency, and standardization shaped how people thought about and transformed food production throughout the twentieth century.

Berry understood these changes better than anyone. He was keenly aware of the seductive and destructive capacities of industrial capitalism. Berry wrote:

> The exploiter is a specialist, an expert; the nurturer is not. The standard of the exploiter is efficiency; the standard of the nurturer is care. The exploiter's goal is money, profit; the nurturer's goal is health—his land's health, his own, his family's, his community's, his country's. Whereas the

> exploiter asks of a piece of land only how much and how quickly it can be made to produce, the nurturer asks a question that is much more complex and difficult: What is its carrying capacity.
>
> ([1977] 2015: 9)

Indeed, for Berry, the conflict between profits and health lay at the heart of an agricultural system under assault.

Berry said very little about insects in *The Unsettling of America*. Yet, when they appeared in the text (as in the epigraph), Berry saw them as symbols of conflict and change. For the exploiters, insects were simply impediments to the productive capacities of industrial agriculture. For nurturers, insects were important contributors to a vibrant and ecologically sophisticated agricultural system. Berry, however, was not the only observer of this tension.

Throughout the twentieth century and across the cultural landscape, insects symbolized both progress and destruction. For many, insects, because they threatened or destroyed crops, became villains of a story that pitted human progress against the hazards of the natural world. For others, insects, while sometimes a nuance, were necessary allies in the production of food, particularly pollinators like bees, flies, butterflies, wasps, and moths. Throughout the century, writers, musicians, scientists, and illustrators shaped the insect story, leaving a set of artifacts that reveal a contested history that is complex and ever changing. This chapter sheds light on that story.

Seen from an American perspective, moreover, insects (and the control of them) became an important part of defining America's political and technological power. As the political and economic influence of the United States expanded across the globe during the century, the quest to control, eliminate, or even eradicate insects served to underscore American scientific and technology proficiency. In this regard, American scientists, policymakers, and cultural producers ascribed insects with greater cultural and political influence during the tumultuous geopolitical transformations of the twentieth century.

Similarly, new technological approaches to food production further altered the political *and* cultural landscapes within and beyond US borders. Indeed, the political ramifications of American agricultural technologies grew precipitously throughout the twentieth century amid fears of over-population and global famines. Set against the tumult of world wars and the Cold War, American agricultural scientists, development experts, and policymakers embarked on missions to reshape agricultural production around the world as a means to garner favor and alleviate famine. In this regard, historian Nick Cullather remarked, "The problem of hunger was one Americans felt uniquely qualified to address" (2010: 70).

Berry's critique on "modern" farming, therefore, did not only resonate within the United States, but across the globe as farmers, agricultural scientists, non-governmental agencies, and governments struggled with the cost, burden, and consequences of industrial agriculture and global famines.

Because global politics framed many questions about food production and insect control in the twentieth century, the cultural impact of insects and food remain deeply contested. Indeed, rather than exploring the culture of insects and food in isolation, this chapter integrates cultural production and meaning within the political and geopolitical realities of the period. It also examines the intersections between science, technology, and culture, suggesting that science has never been an apolitical way of thinking or seeing. When it came to insect control and food production, particularly during the years of the Cold War, American political ambitions were deeply embedded within the very foundations of scientific discovery and technological innovation. The cultural impact of insects and food, therefore, remained central to some of the most important and complex political issues of the twentieth century and today.

***

As others in this series have suggested, the "insect problem" predated the twentieth century. Indeed, the production of basic grains, vegetables, and fruit have benefited from and have been plagued by insects. The United States is no exception. As American farmers migrated westward and embraced new forms of industrial agriculture—plows, combines, monoculture, and later the gasoline-powered tractor—the ecology of the American farm changed, creating environmental conditions ripe for insect infestation. According to historian James McWilliams, by the late nineteenth century, "American farmers were experiencing unprecedented outbreaks of insect attacks" (2008: 81).

As American settlers converted western grasslands into farms, the insect problem grew. The Colorado potato beetle, the San Jose scale, and the chinch bug were some of the more notorious insects that not only threated crops, but were defined disdainfully as "pests." Derived from the sixteenth-century Latin word pestis, meaning plague, the term pest became more commonplace in the American agricultural lexicon and in scientific literature as the insect problem became more pronounced. In 1847, H.D. Richardson published his book *The Pests of the Farm*; *With Instructions for Their Extirpation*: *A Manual of Directions for the Destruction of Vermin*, which offered means to reduce or eliminate the impact of "quadrupeds, birds, and *insects*" (*original emphasis*, 10). Nearly thirty years later, the noted American entomologist Charles Valentine Riley penned *Potato Pests: Being an Illustrated Account of the Colorado Potato-beetle and Other Insect Foes of the Potato in North America* (1876). Riley would become one of the leading authorities of pest control, being named by Congress to head of the United States Entomological Commission in 1877 (McWilliams 2008: 82–7).

As the battle against insect pests became more pronounced, the science of entomology also became more professionalized. Led by people like Charles Riley and Leland O. Howard, the science and methodology of insect control transformed from a practice of individual curiosity and localized knowledge to a highly bureaucratic, state- and federally-funded discipline. Still, there was much debate among the entomological community about the best methods to control or eliminate the insect problem. For example, Riley believed that using an insect's natural enemy would be the most effective way to deal with problem insects. On the other hand, people like Howard began to seek chemical controls as a viable method for insect control. In time, Howard's vision would come to dominate the debate, leading to a fundamental transformation in how humans waged war against insect pests.

While insect pests threatened food and fiber crops alike, cultural producers created and spread stories about their destructive habits across the nation. At the turn of the century, perhaps no other insect captured the attention of the American public more than the boll weevil. Depicted in song, vilified in popular fiction, and memorialized in public monuments, the destructiveness of the *Anthonomus grandis* resonated widely across the nation. Even though the boll weevil did not damage food crops, the political, ecological, and cultural legacy of the insect shaped how food producers and consumers understood the dangers of insect infestation throughout the century.

By 1903, the boll weevil caused farmers massive losses throughout the Cotton Belt of the American South, destroying 25–75 percent of cotton crops in Texas alone (Giesen 2011: 15). By the 1930s, the boll weevil had moved across the Cotton Belt, not only damaging the economic soul of the South, but as historian James Giesen argued, casting fear across the region. In *Boll Weevil Blues*, Giesen wrote, "it was the *idea* of the boll weevil, more than the physical destruction it wrought, that most profoundly changed the [South]" (2011: xi).

By the 1890s, musicians and tenants farmers began singing about the devastation wrought by the boll weevil. "The Boll Weevil" became a staple of the Southern blues, with each musician revising and reinterpreting certain aspects of the song. Later versions of the song reflected the emotional and economic impact of the boll weevil infestation—destroyed crops, financial ruin, broken lives. For example, the folk singer Lead Belly recorded a version of "The Boll Weevil" in the early 1930s that captured the plight of tenant farmers, singing:

> Now the farmer, he said to the merchant,
> I never made but one bale, before I'll let you have that last one,

I will suffer and die in jail,
I will have a home, I will have a home.

During the 1920s and 1930s, the boll weevil remained an essential character for southern blues artists. In 1924, Fiddlin' John Carson sang the "Dixie Boll Weevil." Charley Patton recorded "Mississippi Boll Weevil Blues" in 1929 (Peterson 2007). Five years later, Charles Griffin performed the "Boll Weevil Rag." The legendry ethnomusicologist John Lomax recorded Willie Williams's "Boll Weevil's Been Here" in 1936.

Then in 1939, singer and film actor Tex Ritter performed a version of the "The Boll Weevil" with character actor Mantan Moreland in the film *The Rider of the Frontier*. After the Second World War, the folk renaissance of the 1950s offered new opportunities to reinterpret "The Boll Weevil." The Weavers, the Coachman, and Burl Ives recorded versions of the song, for example. And in July 1961, Brook Benton's version of the blues/folk standard reached number two on the pop charts. More recently, the retro-rock duo The White Stripes often closed their live shows with a cover of "The Boll Weevil" during the early 2000s. While no longer a threat to South cotton farmers, the boll weevil remains central to the cultural legacy of the American folk/blues tradition.

Indeed, the cultural impact of the boll weevil has had a lasting impact throughout the American South. In 1919, Enterprise, Alabama erected the

FIGURE 3.1 Monument to the boll weevil, at Enterprise, Coffee County, Alabama, 1919. Photograph courtesy of Smith Collection/Gado/Getty Images.

Boll Weevil monument, commemorating the powerful effect the insect had in the region.

To this day, residents of Enterprise host the Boll Weevil Festival every October to acknowledge the historical legacy of the insect and a cotton economy that has changed dramatically over the past hundred years (Hall 2013).

The story of the boll weevil underscores another important issue—the threat from a foreign insect invader. Indeed, throughout the nineteenth century, American farmers dealt with an ever-increasing array of "foreign" insects: the gypsy moth, Japanese beetle, the Hessian fly, and the European corn borer, for example. The boll weevil, for its part, migrated from the cotton fields in Mexico. These invasive species cast a specter among the American public, signifying the dangers that lay beyond US borders. The rhetoric of the foreign invader, nevertheless, would shape attitudes about insects and agricultural policy throughout the twentieth century, particularly as American political and military interests expanded around the world.

The First World War proved to be a powerful catalyst in this regard. The war ushered the need for increased food production, leading more and more farmers to cultivate wheat on marginal lands. The plains states of Kansas, Colorado, Nebraska, Oklahoma, and Texas, the noted environmental historian Donald Worster noted, "expanded their wheat lands by 13.5 million acres by 1919, mainly by plowing up 11,000,000 acres of native grass" (2004, 1979: 89). While food production soared during the war years, the long-term ecological impact of this agricultural transformation would be pronounced, resulting in the Dust Bowl of the 1930s. Despite the environmental, economic, and emotional toll of this man-made, "natural" disaster, the belief in modern agriculture remained steadfast in the decades to follow.

The First World War also provided new ways of thinking for pest control. The deployment of nerve gas, however horrific and detested, changed the calculus. The entomological community, for example, reimagined the use of chemicals to control and eliminate insect pests, while the rhetoric of war offered a persuasive language to promote chemicals for insect control. "War," historian Edmund Russell wrote, "created a powerful motive and rationale for a huge leap in the scale on which people controlled insects. *Scientifically and technologically*," he continued, "pest control and chemical warfare each created knowledge and tools that the other used to increase the scale on which it pursued its goals" (2001: 3–4).

Within the United States, perhaps the most forceful and influential advocate for the chemical control of insect pests was Leland O. Howard, the chief entomologist in the Department of Agriculture from 1894–1927. Howard embraced the rhetoric of war to promote a more aggressive approach to insect control. In his 1921 presidential address before the American Association to the Advancement of Science, Howard spoke about "The War

Against Insects." In it, Howard redrew the battle lines, claiming, "there is a war, not between human beings, but between all humanity and certain forces that are arrayed against it." In sounding the alarm, Howard believed that the insect threat was "pressing and immediate," arguing that entomologists should use the "chemical means of war, by spraying our crops with chemical compounds" (Howard 1921).

For chemical manufacturers, this approach to insect control was fortuitous. The rhetoric of war opened an important cultural space that made selling agricultural chemicals consequential to the public at large. Moreover, with the rise of mass culture during the 1920s and 1930s, chemical producers like DuPont, Dow, and Standard Oil began marketing chemicals directly to consumers that reinforced the battle between humans and insects.

For example, in the late 1920s, Standard Oil promoted its insecticide called Flit, hiring Theodor Geisel, better known as Dr. Seuss, to draw cartoon advertisements. According to Edmund Russell, "Flit (and the military metaphors) captured so much of the market that the popular name for the pump-action sprayers (made by any of several companies) was Flitgun" (2001: 86). DuPont, on the other hand, not only created its own corporate magazine—*DuPont Magazine*—but produced "educational advertising," which helped consumers better understand the utility of their products (Junger 2002). They also stoked fears of insect infestation. In a 1938 ad for the fumigant Hydro-Cy, for example, DuPont proclaimed to California fruit growers, "FUMIGATION WILL CONTROL THE RED SCALE MENACE," while casting the insect as "hooded, snarling, simian creature" (Stackman 2005: 78).

Indeed, the imagery of a dangerous insect pest captured the rhetoric of war and became entrenched within the entomological community. Aided by industrial chemists who enthusiastically worked to find new ways to help farmers control or eliminate pests, entomologists pursued an all-out assault on insects. Increasingly, entomologists did not just talk about controlling insects, but eradicating them.

From a public health perspective, the concept of eradication gained traction in the 1910s and 1920s as American health officials sought ways to reduce the threat of tropical diseases like malaria and yellow fever. Armed with chemicals like Paris green, public health officials believed they could eradicate disease from areas once considered inhospitable to human development. From Panama to the jungles of Brazil, people like American physician William Gorgas, the British-born sanitary engineer Joseph LePrince, and noted public health expert Fred Soper reimagined the relationship between humans, tropical diseases, and insects. Over time, and with the assistance of new chemical compounds that were more effective insect killers, the concept of disease eradication became a single-minded campaign to eradicate insects across the globe (Kinkela 2011: ch. 4).

The concept of eradication was based on the belief that humans could control nature with scientific efficiency. And with the proliferation of incredible technical and chemical innovations in the early twentieth century, insect eradication not only seemed possible, but attainable. Indeed, science chartered new approaches to solving complex human problems that made the world "modern."

Nowhere was this belief on display more intentionally than at the 1939 World's Fair held in New York City. General Motors' exhibit, Futurama, captured the celebration of applied science and technology. During the fair, fifty million visitors entered the exhibit hall to discover a diorama of a new world, where, according to Futurama's industrial designer Norman Bel Geddes, it represents "almost every type of terrain in America and illustrat[es] how a motorway system may be laid down over the entire country" (1940: ch. 1).

The companion film, *To New Horizons* (1940), took viewers on a tour of this imagined world, showing the transformation of rural and urban landscapes. Early in the film, viewers see a "modern farm," which operates under "greater security," because "science and research have helped ... to control many of the risks of Agriculture." "Fruit trees," the narrator proclaims, "bear abundantly under individual glass housing," while "orchards are protected against disease and insects." In GM's vision of the future, "physics and chemistry have joined hands with the farmer in helpful friendship" (GM 1940). Indeed, just like the automobile world GM envisioned, with ribbons of highways connecting disparate parts of the United States, the techno-scientific agricultural system they also promoted became an indispensable part of the post-Second World War world.

If the First World War marked a transformative moment defining human/insect relations, then the Second World War strengthened the belief in science and technology to eliminate the insect problem. Scientific innovations from radar to the atomic bomb circumscribed the "good war" with the unyielding trust in the future perfected by technical science. The deployment of a new chemical weapon, dichloro-diphenyl-trichloroethane, or DDT, opened up new possibilities for imagining a world without insect pests. DDT proved to be a powerful chemical technology, killing lice and mosquitoes with remarkable efficacy.

After the war, DDT emerged as the most widely used pesticide in the United States, leading farmers, housekeepers, and public health officials to celebrate the life-changing impact of industrial science. Hailed as a "miracle," DDT changed the calculus of insect control. Farmers, in particular, embraced the new chemical. "No other insecticide," historian Frederick Davis argued, "killed such a broad spectrum of insects without damaging the crops it was protecting" (2014: 91). Indeed, DDT killed insects indiscriminately. Millions and millions of insect pests surrendered to the killing power of new pesticide, but so did

FIGURE 3.2 Shoot to Kill! Poster by Morley. Photograph courtesy of Swim Ink 2, LLC/CORBIS/Corbis via Getty Images.

insects beneficial to agricultural production, like bees and butterflies. Moreover, DDT was far less toxic to people than older metals-based compounds like Paris green. Over the years, DDT's broad-spectrum abilities would become both a blessing and a curse.

Industrial scientists soon brought other chemical pesticides to market. Broad-spectrum pesticides like aldrin, dieldrin, and endrin expanded the chemical options farmers had to control insect populations. Like DDT, these chlorinated hydrocarbons, or persistent pesticides, offered long-lasting protection against insect pests. Somewhat differently, a group of organophosphate pesticides entered the market. Parathion, a compound developed by the German chemical company I.G. Farben during the 1930s, functioned as an extremely effective insect killer, but at nearly twenty times more toxic than DDT posed great risks to humans. And in 1951, American Cyanamid introduced a less-toxic compound malathion. Taken together, the proliferation of chemical pesticides in the postwar period transformed modern agriculture. Indeed, by the 1950s, American farms were awash with chemical pesticides.

While American farmers embraced the new chemical terrain, the underpinnings of its agricultural revolution were being introduced elsewhere. Led by the Rockefeller Foundation, the seeds of "modern" farming were being sown in Mexico. Established in 1943, the Mexican Agricultural Program (MAP) was an attempt to modernize inefficient and "backward" farming. Not surprisingly, insect control and chemical technologies were central to the Foundation's agricultural development project, as were plant hybridization, farm machinery, and specialized knowledge (Olsson 2017: 129–58).

Over time, the Foundation unleashed its agricultural revolution on a worldwide scale, which would be hailed as "The Green Revolution." Led by the brash and energetic agronomist, Norman Borlaug, the Foundation promoted its agricultural vision in Turkey, Iran, Egypt, Pakistan, and India, transforming once dormant lands into productive agricultural landscapes. Borlaug's highly technocratic approach to food production increased agricultural yields

FIGURE 3.3 Nobel Prize-winning biologist Dr. Norman Borlaug holding up stalks of his specifically crossbred wheat, designed to be more disease resistant to produce maximum yields. Photograph courtesy of Keystone-France/Gamma-Rapho via Getty Images.

worldwide, but also intensified the global consumption of chemical fertilizers and pesticides.

Within a decade, the Green Revolution transformed traditional forms of food production into high-yield, capital intensive agriculture. Yet it was not without its critics, many who believed that large-scale agriculture pushed local peoples off their lands and into a system of migrant or dependent labor (Shiva 1991; Cotter 2002). Nevertheless, because of mounting concerns about "famines" and "over-population," particularly within the newly constituted "Third World," many praised the Green Revolution for its humanitarian ambitions. For his part, Borlaug was awarded the Noble Peace Prize in 1970 for his unwavering commitment to transform food production in the developing world.

Geopolitically, the Green Revolution served America's Cold War ambitions as well. It confirmed the nation's generosity and technical skills for a global audience to see, while undermining Soviet influence in strategically significant countries, like India, Pakistan, and Turkey (Cullather 2010). What's more, it connected global food production to the divided politics of the Cold War during the 1950s and 1960s. Food became political, as did the use of pesticides to eliminate insects.

Within the United States, the rhetoric of a divided world—the ever-present threat of communism, for example—became enmeshed in insect control strategies for the public. The rhetoric of domestic consumption of pesticides relied predominately on military language as advertisers and producers marketing the newest "bug bomb" to housewives and calling for "total war" against insect pests. Insects were a "menace" to be eradicated. They were defined as a "horde" or "foreign invader" that sought to undermine the productive capacities of industrial capitalism and the American way of life.

During the Cold War, cultural producers carefully constructed narratives of insects that aligned with the political realities of the world. From advertisers to filmmakers, insects appeared increasingly in America's cultural consciousness as ruthless, destructive enemies of the state. "The sense of public fear of destructive insects, stoked by entomologists, government officials, agricultural interests, and the pesticide industry," historian William Tsutsui wrote, "reached a fever pitch in the 1950s, at the very same time that giant bugs were swarming over movie screens across America" (2007: 247).

Indeed, in 1954, Warner Brothers released *Them!*, a science fiction film about giant mutated ants that threatened the city of Los Angeles.

Years later, American filmgoers were treated to such films as *The Deadly Mantis* (1957), *The Black Scorpion* (1957), *The Fly* (1958), and *Beginning of the End* (1958). This last movie told a story about a horde of giant locusts that ravaged the farmlands of southern Illinois and endangered residents of Chicago. The film captured the tenor and terror of the times, but also

FIGURE 3.4 A poster for Gordon Douglas's 1954 horror film *Them!* Photograph courtesy of Movie Poster Image Art/Getty Images.

framed the human-insect relationship in historical context. In one scene, the film's fictional entomologist, Dr. Ed Wainwright (played by Peter Graves), proclaimed:

> We've been plagued by locusts since Biblical times. We've tried various means of combating them. As a matter of fact, in our own country, the early settlers of the Massachusetts Bay Colony armed themselves with bundles of brush and drove millions of locusts into the sea. Now today, despite the fact that we've developed powerful insecticides, the locust still inflicts damages to the tune of 25 million dollars in the United States alone. California, Colorado, Texas .... Even this small locust will attack a man. It has two powerful jaws that are edged with sharp teeth. It will kill other insects and devour them. If no other insects are available, it becomes a cannibal, turning on its own kind.

> Now you've seen what the locust can do at its normal size, smaller than your thumb. Imagine if you will a locust that's grown bigger than a man and is continuing to grow, some larger than others, but each a deadly killer. I hope you realize we haven't much time.
>
> (Tsutsui 2007: 248)

In each of these films, science was central to the plot. On the one hand, modern science often caused the insect mutation. In *Them!*, exposure to long-term radiation created the conditions for ants to grow to enormous size. In *Beginning of the End*, tests conducted by the United States Department of Agriculture of radioactive fertilizers led locusts to grow to the size of a bus. And perhaps in the most famous monster movie of the period, *Godzilla*, fallout from a nuclear test produced the gargantuan menace that rose from the Pacific Ocean. Atomic science endangered the human world by making insects monsters.

Yet at the same time, these films depicted scientists as the heroic figures. Through guile and ingenuity, the dangers of modern science could overcome its own shortcoming. In this regard, the insect/horror films of the 1950s often grappled with the complexity of science in an ever-changing political and scientific landscape. The proliferation of chemical pesticides to control insect populations during the postwar period both reinforced and exposed the power of science to recast the natural world for human consumption. According to historian William Tsutsui, insect films "reflected a widespread unease about insect infestation and humankind's ability to control it during the 1950s and early 1960s" (2007: 246).

Cultural representations of science often played on this paradox. They were both villain and hero. Still, from entomologists to agricultural scientists, along with much of the scientific community for that matter, the political and cultural landscape in which they operated celebrated scientific knowledge and innovation above all else. And for most Americans, scientists were experts, not villains in some dangerous plot to destroy the world. Although this would change in the decades to follow, during the height of the Cold War, and because of their expertise and the complex geopolitical world in which they lived, scientists were able to shape public opinion and political decision making in profound ways.

The case of *Solenopsis invicta*, or the imported red fire ant, best illustrates the power and danger of scientific authority, particularly regarding the concept of insect eradication. A native species from South American, the black fire ant (*Solenopsis richteri*) entered the United States accidentally on a shipping vessel through port Mobile, Alabama in 1930. Shortly thereafter, the red fire ant arrived in southern ports and quickly overran its South American counterpart. Over time, the red fire ant spread throughout the South, expanding its range to more than twenty million acres by 1957 (Buhs 2004: 23).

Fire ants are mound builders, which, in some cases, "grew so large and so densely that from a distance looked like flocks of sheep" (Buhs 2004: 22). For farmers in the American South, many of whom embraced the technological foundations of modern agriculture of the postwar period, the fire ant became an existential threat and a pest. "The government," one Southern farmer commended, "should be building as big a defense against the fire ants as they are against the Russians" (qtd. in Buhs 2004: 4). The connection between the Cold War and insect invasion could not be made more directly.

In 1957, at this height of the big bug movies, the Plant Pest Control Division of the USDA launched an eradication campaign to wipe out *Solenopsis invicta* with chemical pesticides. To be sure, the cultural representations of insects clearly shaped the control strategies of government entomologists. Historian Pete Daniel remarked, "Although fire ants were not economics pests—that is, they did not seriously interfere with farming—[USDA Agricultural Research Service] publicists and other supporters transmogrified the insect into a monster that devoured quail, calves, crops, and the tender flesh of children" (2005: 57). Indeed, the chomping mandibles of the mutated ants from *Them!* undoubtedly fit into the USDA'S narrative that described fire ants as a menace and threat.

Moreover, the zeal with which government entomologists attacked the fire ant was only matched by their hubris. In their quest to eradicate the fire ant, USDA experts deployed crop-dusters to spray millions of gallons of the pesticides throughout the South. Armed with chemicals like heptachlor, dieldrin, aldrin, and Mirex, the war against the insect intensified each year. The power of science and technology became an intoxicating elixir for many in their quest to eradicate insects once and for all.

Somewhat surprisingly, crop-dusters became part of American agricultural *and* cultural landscape. In his 1959 critically acclaimed film *North by Northwest*, Alfred Hitchcock reimagined the aerial assault of insects in the famous crop-duster scene.

In it, a menacing crop-duster attacks the well-dressed Cary Grant. The pilot first shoots at his intended victim with a convention weapons—a gun—but misses. Grant quickly finds shelter in a cornfield. The pilot then unleashes a fog of chemical pesticides that not only chokes the victim, but as if fulfilling the eradicator's prophecy, forces him from the safe haven of the corn. Grant eventually escapes as the zealous pilot mishandles the crop-duster, crashing it into an oncoming oil truck.

The scene, however, resonated across the cultural landscape. Hitchcock, himself, was often asked about the scene. In a 1965 interview, Hitchcock, sounding remarkably like an USDA fire ant eradicator, proclaimed, "One rule I have, one very essential [rule], if you have a crop-duster, you must let it dust" (*Hitchcock s'explique* 1965). And let it dust Hitchcock and the USDA did. Over

FIGURE 3.5 Cary Grant fleeing from a plane in a scene from the film *North By Northwest*, 1959. Photograph courtesy of Metro-Goldwyn-Mayer/Getty Images.

the span of twenty years, USDA eradicators sprayed nearly fifty-seven million acres of farmland to eradicate the *Solenopsis invicta* across the South.

While *North by Northwest* would go on be to one of the most culturally significant films of the twentieth century, the USDA's fire ant eradication spray campaign would come to an ignoble and ecologically costly end. Despite the massive spray campaign, the fire ant endured. Moreover, collateral damage from the spray campaign—poisoned fish and foul, for example—reshaped both the southern landscape and people's ideas about eradication. The noted biologist Edward O. Wilson claimed that, "The fire ant control program in the South is the Vietnam of entomology" (Daniel 2005: 129). Indeed, it was a battle that couldn't be won with more powerful technologies and a bombing program on par with Vietnam's "rolling thunder."

Criticism of insect eradication was not an afterthought. Many biologists and entomologists understood the theoretical underpinning of eradication were unsustainable. But given the widespread institutional support of the initiative, whether that was from the USDA, the Rockefeller Foundation, or the World Health Organization, who launched a global malaria eradication campaign in

1955, eradication became the predominate lens in which to understand insects within the "modern" world (Kinkela 2011: ch. 4).

In 1962, Rachel Carson, a well-known science writer, challenged this perspective. Her landmark book *Silent Spring* awoke readers to the dangers of how chemical pesticides were being used and the short-sightedness of experts who believed they could control nature. "The crusade to create a chemically sterile, insect-free world," she wrote, "seems to have engendered a fanatic zeal on the part of many specialists and most of the so-called control agencies" (1962: 12). For Carson, the so-called "experts" and institutions they built were incapable of seeing insects as anything but a problem. And their desire to control nature with chemical pesticides was "conceived in arrogance" (297).

Silent Spring offered an alternative vision of insects. For her, they were "a group of extraordinarily varied and adaptable beings," and an essential part of what she called "the balance of nature" (9, 114). Moreover, Carson's articulation of the balance of nature introduced her readers to the science of ecology. She encouraged readers to understand the complex interdependency of the natural world. "Only by taking account of such life forces and by cautiously seeking to guide them into channels favorable to ourselves," Carson wrote, "can we hope to achieve a reasonable accommodation between the insect hordes and ourselves" (296).

Carson did not blame farmers, however. Rather she criticized techno-science and institutional thinking. She condemned the "devotion to immense acreages to a single crop," claiming that this "does not take advantage of the principles by which nature works; it is agriculture as an engineer might conceive it to be" (10). For Carson, chemical producers, agricultural "experts," and state and federal agencies collectively promoted a vision of food production that ran counter to the ecological complexity of the nature world. And while her book became an instant bestseller and remains one of the critical texts on the environment to this day, the evolution she espoused has been slow to materialize.

Within the United States, *Silent Spring* led to changes in environmental policy at the federal level. In 1972, for instance, the newly created Environmental Protection Agency (EPA) cancelled the registration of DDT. Shortly thereafter, the EPA banned the use of other chlorinated hydrocarbons as well. These decisions represented a triumph of Carson's position. Over time, similar prohibitions would be made worldwide. Yet, within the context of industrial agriculture, insects remained a problem and the solution remained chemical pesticides, which became far more dangerous to those who plant, cultivate, and harvest food crops. "Contrary to Carson's clarion call for reduction in the use of all insecticides," historian Frederick Davis noted, "the ban of DDT ... initiated a risk-risk trade-off in which agribusiness replaced DDT [and other chlorinated hydrocarbons] with highly toxic organophosphates, like parathion, that threaten the welfare of human and wildlife" (2014: 223). In fact, organophosphates, like parathion and malathion, remain among the most widely used pesticides around the world to this day.

Other critics of industrial agriculture emerged as well. Influenced by the British botanist Sir Albert Howard and the indomitable American entrepreneur J.I. Rodale, American backyard farmers began to cultivate food organically. Howard's books, *An Agricultural Testament* (1940) and *The Soil and Health: A Study of Organic Agriculture* (1947), sprang from his exploration of local agricultural methods while serving as an agricultural advisor in colonial India. Rodale, for his part, began publishing the magazine *Organic Farming and Gardening* in 1940, claiming that growing food should be done "Nature's way." Soon after, Rodale published a series of books promoting organic farming, including *Pay Dirt: Farming and Gardening with Composts* (1945), *The Organic Front* (1948), and *the Encyclopedia of Organic Farming* (1959), which remains in publication to this day.

For Rodale, a deep appreciation of the complexity of the natural world lay at the heart of his critique. "Farming," Rodale proclaimed in *The Organic Front*, "has been too much reduced to a chemical formula by unimaginative men" (qtd. in Jundt 2014: 50). Hailed as the father of organic farming in the United States, Rodale would have a distinguished career until his untimely death in 1971. Nevertheless, his work continues through the Rodale Institute (est. 1947) and Rodale, Inc. (est. 1930), the last of which publishes a variety of magazines and books on health and wellness, including *Organic Gardening* and *Men's Health*.

The rise of the organic farming movement, which aligned unequivocally with the political and cultural impact of *Silent Spring*, created a ground swell of support for an agricultural system devoid of industrial chemicals. "Since the 1940s," historian Robin O'Sullivan wrote, "the concepts of holism, health, natural cycles, and balance have been integral to the organic ethos" (O'Sullivan 2015: 12). It is an ethos that at times has shunned empirical evidence in favor of spiritual connection to land and labor. Indeed the movement, which captured greater numbers during the 1970s and 1980s, was often derided as being too idealistic and incongruent with the geopolitical realities of the world. In 1971, for example, Earl Butz, the Secretary of Agriculture stated, "before we go back to organic agriculture, somebody is going to have to decide what 50 million people we are going to let starve" (O'Sullivan 2015: 1). This position held a powerful position at the time, particularly since biologists like Paul Ehrlich sounded the alarm about human population growth and the likelihood of extensive famines in his 1968 bestseller *The Population Bomb*. Never far from politics, the contest between different forms of food production in the United States and elsewhere became pronounced and political during the 1960 and 1970s (Kohn 1989).

Not surprisingly, insects were at the center of this emerging debate. And while the organic movement espoused reducing insect pests naturally, new strategies, techniques, and approaches to support industrial agriculture's quest to eliminate or reduce harmful insects from crop production appeared. Insect sterilization, for example, offered an alternative to chemical pesticide for certain insect species. More holistically, Integrated Pest Management (IPM) sought to combine biological and chemical strategies as a way to reduce crop losses from insect pests (Kogan 1998). And the introduction of Genetically Modified Organisms (GMOs) created a ground-breaking, yet highly contested way to reduce sprayable pesticides for crop production.

In the 1950s, USDA entomologist Edward Knipling developed a technique to sterilize male screwworms, a flesh-eating insect that damaged livestock populations. Knipling, who also conducted research on the efficacy of DDT during the Second World War, promoted sterilization as a viable means of pest control, leading to what the entomology community called Total Population Management (TPM) (Knipling 1963, 1964). Later in his career, Knipling became a proponent of IPM, which relies on "natural control factors to the maximum extent before resorting to the application of insecticides." Yet, as Knipling and others understood, "natural controls – as vital as they are – do not provide the protection needed for a wide range of persistent insect pests" (Office of Technological Assessment 1995: 60).

The same is true with GMOs. During the late 1980s, Monsanto, the multinational biotechnology corporation, introduced glyphosate-tolerant soybean (or what became known as RoundUp Ready soybean) for weed control. In the 1990s, researchers combined *Bacillus thuringiensis*, a naturally occurring soil bacterium with insecticidal properties to certain feeding insect larvae, with corn to produce Bt corn, a GMO designed to prevent crop damage by the European corn borer (*Ostrinia nubilalis*). The successful modification of corn led to other Bt modified crops, including potato, soybean, canola, and tomato. By producing seeds based on recombinant DNA technology, Monsanto (and other agricultural multinationals for that matter) believed their approach would reduce the need and use of sprayable pesticides (Nelson 2001; Tauger 2011: 138–79). The evidence suggests that chemical pesticide use has declined somewhat with the introduction of genetic-engineering technologies. However, the global pesticide market continues to expand in the early twenty-first century (Peshin and Zhang 2014: 1–46).

The recent biotechnology revolution has profoundly transformed the cultural and political implications of food production. Whereas non-government agencies like the Rockefeller Foundation spearheaded the Green Revolution of the 1950s and 1960s, the drivers of this recent agricultural transformation almost exclusively have been multinational corporations like Monsanto (US), Bayer (Germany), Syngenta (Switzerland), DuPont (US), and Dow (US). Driven by profits, patent rights, and industrial trade secrets, the process of industrial food production (and insect control) are intrinsically bound to the principles of privatization and the free market. GMOs are not simply another manifestation of agricultural science, but have become a staple of the neoliberal upheaval of the past fifty years.

Indeed, ownership of seed technology has made agricultural production fraught with uncertainty and fiscally dangerous to some farmers. Farmers used to worry about insect pests and poor weather. Now, biotechnology firms have sought legal action against farmers using GMO seeds without paying royalties. And while the public rhetoric about Monsanto suing small farmers over crosspollination—that is non-GMOs crops being pollinated by GMO varieties—is vastly overstated, the economics and politics of food production in the twenty-first century have been fundamentally altered.

The fractured politics and economic decline of the 1970s, which many historians argue created the conditions for neoliberalism to emerge, also witnessed a re-emergence of insect films. Film and television scriptwriters capitalized on a new insect threat—killer bees. This highly aggressive bee species did not "invade" the United States until the 1990s, yet the story of the bee's migration from Brazil to Mexico captured the public's attention (Hatcher and Battey 2011: ch. 4).

In 1957, Brazilian honey producers, who previously relied on the European honey bee (*Apis mellifera mellifera*), imported the African honey bee (*Apis mellifera scutellata*), only to discover that the bee protected its hive with much ferocity. As the bee migrated through South and Central America, the story of garnered much attention. On August 14, 1965, Reuters reported that "Killer Bees Plague Brazil." In April 1976, *National Geographic* reported on "Those Fiery Brazilian Bees."

Amid news reports of the impeding invasion of African Killer Bees, cultural producers played into the public's fear. And, once again, film and television producers depicted insects as a menace to humans, despite being only an existential threat to Americans. Films like *Killer Bees* (1974), *Bug* (1975), *Savage Bees* (1976), *Empire of the Ants* (1977), *It Happened at Lakewood Manor* (alternative title, *Ants*, 1977), *Swarm* (1978), and *The Bees* (1978) all added to growing fear of insects and the killer bee invasion. Yet, unlike the big bug films of the 1950s, in which insects grew to enormous size, the films of the 1970s emphasized the swarming nature of insect population. It wasn't their size, but the sheer number.

Moreover, these films played off familiar tropes about the power and limits of science. In *Swarm*, the bees could not be controlled or eradicated with chemical pesticides. When the lead scientist developed "pellets" to poison the swarm, which, being the post-*Silent Spring* era, would not "harm ... the population or plant life," the bees simply outsmarted their human counterparts. Only by relying on sound and the reproductive desires of the killer bees was the swarm eventually vanquished.

Despite being acknowledged as one of the worst films ever made, *Swarm* attempted to differentiate "good" bees from its invasive killer relatives. As the film concluded, the film's director Irwin Allen added a final disclaimer before the credits rolled. He reminded the audience that "The African killer bee portrayed in this film bears absolutely no relationship to the industrious, hardworking American honey bee to which we are indebted for pollinating vital crops that feed our nation" (*Swarm*, 1978).

Not to be outdone, even Saturday Night Live (SNL) captured this unique moment with their 1976 skit, "The Killer Bees: Home Invasion." In it, John Belushi, Elliot Gould, Dan Aykroyd, and Garrett Morris attack a suburban home demanding pollen from a middle-class American couple played by Gilda Ratner and Chevy Chase. Playing stereotypical "Mexican banditos," one of the killer bees demanded, "Your pollen or your wife, señor!" After discovering that the suburban couple did not have pollen, the lead bee stated, "Let the woman go. One more killing would not put pollen in the mouths of our starving children" (SNL 1976).

Comedy aside, "The Killer Bee" sketch played on a number of key themes pertaining to the cultural history of insects and food. First, the depiction of the bees as Mexican banditos reinforces the idea of outside invasion. Indeed, many of the pest control strategies developed in the twentieth century were aimed at invasive species—the boll weevil, the fire ant, the European corn borer, for example. Second, the idea of potential violence and the desire to feed "starving children" also underlined the logic of the Green Revolution amid fears of overpopulation and famine. By connecting the geopolitical complexity of food production and protection of the home front, SNL reframed the killer bee story into a familiar narrative that underscored the history of humans and insects during the twentieth century and one that late-night audiences understood.

Three decades later, media reports sounded the alarm about bee populations once again. In the early twenty-first century, beekeepers and entomologists discovered that honey bees had simply vanished. Bee colonies that were seemingly healthy and prosperous collapsed, leading many to wonder why. During the winters of 2006–7 and 2007–8, for example, honey bee populations declined by nearly 32 percent and 36 percent, respectively (Johnson, R 2010). In 2007, the *New York Times* reported on a "mystery disease" that threated bee colonies (February 12, 2007). In another story, the *Times* proclaimed, "Bees Vanish, and Scientists Race for Reasons" (April 24, 2007). And, to reinforce the paradox of this latest insect problem, *The New Yorker* simply stated "STUNG: Where have all the bees gone" (August 6, 2007). Over the past ten years, bee colonies have also been affected in Europe and Asia.

In what became known as colony collapse disorder (CCD), the press *and* the public grew alarmed not because there were too many insects, but too few (Jacobsen 2008; Schacker 2008; Benjamin and McCallum 2009). Indeed, the bee colony collapse shifted the public discourse about insects. Rather than seeing them as a menace or threat that needed to be controlled, the colony collapse reminded the public that bees served an invaluable and necessary part of modern food production. Some reports claim, "bees contribute more than $15 billion to U.S. crop production" (Holland 2013). Indeed, large-scale agriculture required managed bee colonies to be brought to fields to pollinate fruit and nut trees as well as alfalfa, strawberry, and other plant species. "We can't count on wild pollinators," May Berenbaum, a professor of entomology at the University of Illinois claimed at the height of the CCD, "because we've so altered the landscape that many are no longer viable" (Kolbert 2007).

While honey bee colonies have rebounded somewhat since CCD first appeared, the threat remains real. A decade of research to find the cause has yielded no clear answers. According to the scientific literature, colony collapse can be attributed to a combination of factors, including the use of neonicotinoids, a new class of pesticides that may increase bees' vulnerability to other diseases, the *Varroa* mite, viral infections, climate change, and habitat destruction. Taken together, the health of honey bee populations around the world remains a concern for many people.

In 2007, as the scientific community and the public at large fretted about the extent of the bee colony collapse, DreamWorks released *Bee Movie*. Written by the comedian Jerry Seinfeld, the film captures the intertwined histories of bees and humans in the production and consumption of honey. Through a series of convoluted encounters, the bees win the legal right to retain all the honey they produce, which leads to an oversupply of honey and eventual ecological catastrophe. Because bees had an abundance of food, they no longer pollinated plants, causing a post-apocalyptic "natural" landscape to emerge. "It's not just flowers," Vanessa, the lead human character proclaimed, "Fruits, vegetables, they all need bees." While aimed at children, *Bee Movie* celebrated the reciprocal relationship between honey bees and humans. Only by working together, the film asserts, could some sort of ecological equilibrium be achieved.

This storyline is far from the horror films of the past. *Bee Movie* does not vilify insects. Nor do such animated films like *Antz* (1997), *A Bug's Life* (1998), and *James and the Giant Peach* (1997), which tend to cast insects in a positive light; the menacing grasshoppers from *A Bug's Life* notwithstanding. Of course, there is a long history of popular, anthropomorphic insect characters in animated films: Jiminy Cricket in *Pinocchio* (1940), Absolem the Caterpillar in *Alice in Wonderland* (1951), and while technically an arachnid, Charlotte in *Charlotte's Web* (1973, 2006). These depictions of the insect world both challenge and reinforce our cultural understanding of insects because they complicate a specific cultural, scientific, and political narrative that has prevailed throughout the twentieth century, namely that insects are a menace to the human world. Seen through the lens of animated storytelling, these characters tell us that insects are vital to the human world. It is a story not often told.

***

Throughout the twentieth century, insects have caused humans a lot of grief, concern, and money. They have been vilified as harmful pests capable of inflicting enormous pain and suffering. They spread disease, ravage crops, and disrupt economic development. Culturally, the story of insects and food in the twentieth century has emphasized these negative traits. In song and film, insects have been branded as the cause of human misery.

At the same time, humans have celebrated their technical achievements designed to annihilate insects. The proliferation of chemical pesticides and the promise of an insect-free world produced grand schemes to vanquish the ancient foe. Eradication became the predominate ideology that shaped human/insect interaction during the mid-twentieth century. And modern science and technology were the means to finally eradicate insect pests. As we have seen, the logic of eradication proved to be faulty. Aided by proponents of organic farming and a more rational approach to insect control, the paradigm that defines modern agriculture has shifted, but the underlying premise of industrial agriculture remains extremely powerful.

As for insects, they will always be a part of the human world. In fact, insects will continue to thrive even if (or when) humans cease to exist on planet Earth. Yet, our complicated and contentious relationship with insects persists. No longer does the rhetoric of eradication or the unwavering belief in science and technology exist, although there are those who continue to espouse such views.

Over time, however, we have seen the rise of organic food production and non-chemical methods of insect control. From its humble beginnings in the 1940s to the proliferation of farmers markets, supermarkets like Whole Foods, and local food movements, different forms of food production have shifted the political, ecological, and cultural landscapes.

Returning to *The Unsettling of America*, Wendell Berry envisioned a world in which people shared "ownership of the land and thus be bound to it by economic interests, by the investment in love and work, by family loyalty, by memory and tradition." In doing so, Berry reimagined a neo-Jefferson agrarian democracy based on an "economy of necessities rather than an economy based on anxiety, fantasy, luxury, and idle wishing." More importantly, perhaps, Berry believed this reawakening of American landownership and labor would create an "agriculture based upon intensive work, local energies, care, and long-living communities – that is, to state the matter from a consumer's point of view: a dependable long-term food supply" (1977: 14).

Berry's vision of sustainable food production would not come to fruition. However, writers like Michael Pollan continue to reframe the culture, economics, and politics of food production. In his bestselling 2006 book *The Omnivores Dilemma*, Pollan argues, "A great many of the health and environmental problems created by our food system owe to our attempts to oversimplify nature's complexity." Pollan, a critical observer of the natural world, celebrates the complexity of nature and the harmony between plants, soil, sun, insects, and, in some cases, humans. "Eating," he wrote, "puts us in touch with all that we share with the other animals, and all that sets us apart. It defines us" (2006: Introduction).

So as we grapple with the complexity of food production in our own time, it is important to look at the ways in which humans define and shape the world around them. As I have argued in this chapter, the political, agricultural, and cultural practice and representations of insects and food production have changed over time. And while the trajectory of this narrative seems to suggest a more hopeful and sustainable future, history is never linear. Industrial models of food production and insect control continue to hold a powerful place in our world. However, the significance of storytelling—whether a song about the boll weevil or a film with talking bees, for example—provides important cultural space to challenge existing ideas about insects and the ecology of food production.

CHAPTER FOUR

Insect Products

AMELIA DOLAN

> When we try to pick out anything by itself, we find it hitched to everything else in the universe.
>
> (Muir [1911] 1988: 110)

Over the course of human history, insect products have been used in a myriad of ways. Honey and wax have been harvested from bee hives (*Apis* species) to add sweetness to our meals and provide candlelight. Silkworms (*Bombyx mori*) have provided the raw material for intricate garments, accessories, and tapestries. Resins collected from certain scale insects (Family Kerriidae) have been used in dyes, cosmetics, wood finishes, and even phonograph records. In the modern age (1920s to present), some things have not changed; these classic insect products continue to be valuable, and, in some cases, they have become big business. But in other ways, the modern era brought about a considerable transformation in our understanding and consumption of insect products.

Human innovation, advances in technology, urbanization, globalization, and the rise of commercialism have all affected the value placed on insects and their products in the modern age. To truly appreciate the value of insects and their products in the twentieth and twenty-first centuries, one must recognize several things: 1) advances in technology have led to exponential increases in our understanding of the incredible diversity of insect species and the innovative adaptations they have for survival; 2) the definition of "product" in modern society goes well beyond a simple commodity; and 3) since the early twentieth century, the developed world has experienced a major cultural shift to urbanization, global capitalism, commercialism, and consumerism.

The world has undergone an extraordinary transformation since the early 1900s. Following the Industrial Revolution of the eighteenth and nineteenth centuries, society changed. The replacement of manual labor with machines and the widespread use of fossil fuels to power those machines marked just the beginning of the impacts of industrialization. The Industrial Revolution moved across the globe and its long-term effects included not only urbanization and exponential population growth, but changes in lifestyle, values, food production methods, the quality and quantity of manufactured products, and advances in science, medicine, and technology (McLamb 2011). The Industrial Revolution started in Britain about 1760 and spread to other parts of the world by 1860. In reality, though, the global evolution of industry and innovation (i.e., "the industrial revolution") never really ended. In fact, human knowledge and technological capacity have advanced more rapidly since 1920 than in any other era. Many scientific and technological breakthroughs evolved out of the world wars in the 1910s and 1940s, and in the twenty-first century both products and knowledge continue to be refined at incomprehensible speeds.

Since the beginning of the twentieth century, humans have moved from radios to black and white televisions to high definition flat screens for personal entertainment. Communication media have shifted from the telegraph to the telephone to cell phones then smartphones, and finally to a world of social media in which the average person (in 2011) was inundated with 174 newspapers' worth of data each day (Hilbert and Lopez 2011). The giant, room-sized computing machines that were cutting-edge technology in the 1940s through the 1960s have been refined at exponential speeds so that by the 2010s individuals are able to carry high-speed computers in their pockets. Vehicles of transportation by land, sea, and air have all been transformed to become faster, cheaper, safer, and more energy-efficient so that in the first half of the twenty-first century it is not unrealistic for an individual to travel to any location on Earth. Scientific advancements have allowed for the scrutiny of objects as small as an atom using scanning electron microscopes and the exploration of galaxies light-years away with high-powered telescopes and complex spacecraft on the move throughout our solar system. After the molecular structure of DNA was first determined in the 1950s and the Human Genome Project was completed by 2003, genetic engineering technologies became commonplace in research, biotechnology, agriculture, and medicine. Breakthroughs in chemistry that brought about chemical warfare in the First World War and synthetic pesticides in the 1940s were the foundation of modern chemistry's capacity to produce any number of synthetic products, all meant to make every aspect of human existence faster, easier, and safer.

In the field of entomology, these scientific and technological advancements revolutionized understanding of insects and their products. We have discovered that insects are more numerous, diverse, and valuable than was realized in previous eras. Entomologists are beginning to better understand the natural and genetic histories of this group and how they are adapted to survive in nearly every habitat on Earth. It is now relatively easy for scientists to travel to remote sites across the globe to collect and record all sorts of entomological data. The depth of new understanding is reflected in the literature as the narrative descriptions of insect morphology and behavior commonly published by entomologists through the 1800s has been replaced by highly technical publications involving DNA sequences, deconstructed biochemistry, and advanced statistical analyses.

These scientific and technological insights facilitated a substantial shift in the modern era and an evolving definition of the term "product." In previous ages, the term "product" was limited in scope to tangible *things*—commodities, merchandise, or raw materials (e.g., honey, silk, or dyes). In the modern era, however, the types of products attributable to insects began to expand. *The Economic Times* (2017) defines "product" this way: "A product is the item offered for sale. A product can be a service or an item. It can be physical or in virtual or cyber form." This expanded definition of insect products is perhaps the most unique aspect of understanding insect products and their value in the modern age.

Insect products, in the most basic sense of "product" (i.e., a commodity, a thing), have in many ways become obsolete in the twentieth and twenty-first centuries. Synthetic products (mostly derived from petroleum) have been manufactured to serve the same purpose as insect products, often more effectively and less expensively. Today, paraffin is the most commonly used wax in candle making, displacing beeswax. Shellac, used as varnish and to make phonographic records, has been replaced by a synthetic plastic resin. Silk is still considered a luxury fabric, but its use is also on the decline due to synthetics like polyester and nylon. Honey is still ubiquitous on the shelves of food markets, despite the enormous popularity of cane sugar, beet sugar, and high fructose corn syrup as sweeteners, but honey, especially "local honey," is now marketed as a specialty crop—a healthier alternative to sugar or a natural remedy for a variety of ailments.

At the same time that these classic insect products are becoming obsolete, a new group of insect products has been on the rise. One of these includes synthetic products or systems *inspired* by insects and their adaptations. In this case, the products coming directly from insects seem to merely be ideas. But these ideas are proving to have scientific, economic, and environmental importance. In Harare, Zimbabwe, for example, architect Mick Pearce designed Eastgate Center (the country's largest shopping/office complex) to mimic the cooling system found in African termite mounds. In doing so, he created a building that uses only 10 percent of the energy of a "normal" building of similar size (Tonn 2015). Termite-inspired buildings are just one success in the relatively new field of biomimicry, as scientists use modern technological advances to understand the chemicals, structures, or systems found in nature and apply their discoveries in the innovation of sustainable solutions to human problems.

Beyond the innovative scientific inspiration that emerges from in-depth understanding of the biology, physiology, and ecology of insects, entomologists have developed deeper understanding of the role insects play in the natural world. A key realization that some of the most valuable insect products are the plethora of *services* they provide.

For one, insects help solve crimes. Defined as "the use of insects (or related arthropods) that inhabit decomposing remains to aid legal investigations" (Byrd 1998), forensic entomology has become relatively routine in the modern age, though the use of insects to solve crimes dates to thirteenth-century China (Benecke 2008). When an organism dies, bacteria and fungi begin the decomposition process. These decomposing remains attract several species of flies (most notably blow flies, family Calliphoridae) and beetles (families Silphidae, carrion beetles; Dermestidae, dermestid beetles; and Staphylinidae, rove beetles), which feed on and lay eggs in the carcasses, aiding the process of decomposition. Different stages of decomposition attract specific insect species. Thus, by identifying which species are present in which stage of their life cycles, forensic entomologists are able to estimate time of death.

Since carrion beetles such as the "hide beetle" (*Dermestes maculates*) are known to feed on dead or decaying animal flesh, their services have been used beyond forensic entomology. Colonies are cultivated and bought by museum curators or taxidermists to prepare skeletons (i.e., removing flesh from the bone) for research or display (Bloudoff-Indelicato 2013). A well-established beetle colony can be very efficient, cleaning a large animal skull in just 24–48 hours. Beetles are preferable to other flesh-removing methods since chemicals can cause damage to bones and boiling leaves behind greasy, yellow residues.

Solving crimes is a fascinating service provided to humans by insects, but on a global scale its impact is relatively small. Perhaps the greatest services that insects provide are activities classified as "ecosystem services." This term, "*ecosystem service*," was defined and popularized by the Millennium Ecosystem Assessment published in 2005 as "benefits people obtain from ecosystems." There are four categories of ecosystem services: provisioning, regulating, cultural, and supporting. When humans gather needed resources, such as food, water, or timber from an ecosystem, the ecosystem is providing a *provisioning* service. When an ecosystem is able to stabilize climate, prevent flooding, improve water quality, etc., it is considered a *regulating* service. *Cultural* ecosystem services provide recreational, aesthetic, or spiritual benefits to people. *Supporting* ecosystem services, such as photosynthesis and the formation of soil, are those processes that enable the system to persevere (Hassan et al. 2005). Although this way of looking at the structure and function of complex ecosystems is extremely anthropocentric, it serves its purpose as a lens through which to educate the public and raise awareness, which can lead to funding for research and conservation.

Insects contribute to all four categories of ecosystem services. Insects pollinate, disperse seeds, supply nutrients, and in some cases, provide defense to a wide variety of plants. As primary consumers in most food webs, insects are an essential source of energy to all other trophic levels. As regulators, insects are able to limit the population growth of plants, vertebrates, and other insects, and without certain insect scavengers and detritivores, all landscapes on Earth would be littered with dead animals, plants, and dung. In 2006, four specific ecosystem services (dung burial, pest control, pollination, and wildlife nutrition) were measured and analyzed through the lens of economic value. The analysis of wild insects (managed populations like honey bee hives were excluded from the study) performing just four ecosystem services estimated the economic value of those insects to be at least $57 billion in the United States alone, indicating that the global value of the ecosystem services of all insects is likely to be incredibly high (Losey and Vaughan 2006).

In the mid-twentieth century, the national economies of developed nations began globalizing, driven by advances in technology and communication. As everyday products evolved to become became "better," "faster," and "easier," a cultural shift to commercialism occurred. In the twenty-first century, products are most often discarded and replaced, instead of being repaired. Merchandise is specifically designed to become obsolete in a short amount of time, fueling the habits of a "throw-away" generation (Guiltinan 2009). Meanwhile, online marketplaces make it easy for individuals to buy or sell any product, service, or idea. Insects and their products have not been immune to this commercialization. Any insect, insect product, or product made from insects that can be sold is being sold. With the internet as a marketplace and the availability of social media groups or online blogs to offer "how-to" advice, just about anyone with a little bit of knowledge, creativity, and an entrepreneurial spirit can make money off insects and their products.

In some cases, insects themselves are the product. Honey and bumble bee colonies are marketed to gardeners as must-have pollinators while ladybird beetles, preying mantids, lacewings, and a variety of wasps are sold as garden "beneficials." Insect houses, food, attractants, and lures can also be found at garden stores. Live insects and other related arthropods such as cockroaches, tarantulas, scorpions, and ants are sold as pets, for insectaria, or to schools as part of an engaging elementary science unit of study. A "butterfly release" is marketed as a beautiful way to celebrate a marriage or honor a loved one's passing and tours are sold to visit stops along the famous North American monarch migration route. Insects are also bred and sold for scientific and medical research, pet food, and in many places, for human consumption. Even carrion beetles have been commercialized. A quick internet search can lead any amateur taxidermist to multiple vendors selling beetle adults or larvae for $55 to $450. Those same sites offer a plethora of "how-to" advice on keeping a colony of dermestid beetles and using them to clean skulls or other bones.

Products directly from insects like honey, wax, resins, and dyes continue to be sold as they have been for centuries. But in the second half of the twentieth century, a variety of secondary insect products have joined the market. These secondary products include lotions, cosmetics, waterproofing products, holiday ornaments, herbal remedies, food wraps, pasta, candy, jewelry, framed insect art, and furniture polish. Products indirectly related to insects are also being sold in stores and online. These include soothing audio files of recorded insect sounds, children's books, "how-to" books for hobbyists, toys, and services to identify or remove insects from people's homes.

POLLINATION SERVICES OF BEES IN THE UNITED STATES: A CLOSER LOOK

Perhaps the best example of the intricate interconnectedness of insect services, human innovation, urbanization, and commercialism can be found by taking a closer look at insect pollinators, specifically the pollination services of bees in America in the twentieth and twenty-first centuries.

It is relatively easy to argue that pollination services are the most important of the ecosystems services, and perhaps even the most important insect product. Global food staples (i.e., rice, wheat, corn, and soybean) are wind or self-pollinated, so humans could, in theory, survive without insect pollinators. However, without insect pollination, the human diet would be much less varied and therefore less nutritious. Approximately three-quarters of the world's 240,000 flowering plants rely to some degree on animal pollinators (insects, bats, birds). These plants include bee-pollinated fruit, vegetable, and seed crops and those that provide fiber, fuel, and drugs (National Research Council 2007: 2). Animal feed crops, such as alfalfa and clover, also rely on insect pollinators, making insects a vital part of the beef and dairy industries. Beyond their economic value, insect pollinators make human recreational and cultural activities such as gardening, berry-picking, and wildflower hikes possible.

If the average person in the twenty-first century was asked to share what they know about bees, chances are, all the facts and stories shared would be about honey bees: they live in hives, there is a queen and workers, the hive produces honey and wax, they are pollinators, they can be dangerous when they sting. Because of their famous products and the fact that they have been domesticated, honey bees have become the poster child for all bees. In reality, however, honey bees represent just one, minute subset of a diverse group of insects.

There are more than 17,500 described species of bees (Michener 2007). Of these, the honey bees and stingless honey bees comprise just a few genera in the family Apidae. The vast majority of the bees of the world are small, solitary, wild, native bees. Ecologically, bees can be placed into two groups based on the relative length of the proboscises ("tongues") used to gather nectar from flowering plants. Taxonomically, bees are placed in seven families. These family classifications can be used to make generalizations about the biological activities (nesting, foraging, habitat preference) of different species.

Species in the family Andrenidae are commonly known as mining bees. These bees are solitary species that nest in the ground. Although andrenids could be considered cosmopolitan, most are found in temperate and arid habitats. They tend to be relatively small and many specialize in collecting pollen from only a few genera or species of plants (oligolectic). The family Apidae contains the most well-known bee species: honey bees, bumble bees, orchid bees, and carpenter bees. This is the largest bee family, with nearly six thousand identified species. Biology is diverse among the apids. Honey bees and bumble bees form eusocial colonies, meaning there is one reproductive female and several overlapping generations of non-reproductive adults that cooperatively care for young, maintain the nest, and forage for resources. Carpenter bees are solitary but gregarious and nest in plant material. A few apid species are cleptoparasitic, meaning the female kills the queen of another species and hijacks the nest, using the resources (including pollen, nectar, and even worker bees) to rear her own young. But the majority of apids are solitary species that build nests in ground or tree cavities. Bees in the family Colletidae are often referred to as "plasterer bees" because they line their nests with a secretion that dries to the texture of cellophane. The species in this family are all small, solitary bees and most diversity can be found in South America and Australia. Family Halictidae is composed of small, often-metallic, ground-nesting bees that can be attracted to perspiration. Hence their common name, "sweat bees." Some halictid species are considered at least partially eusocial, some are cleptoparasites, but most are solitary. This is the second largest bee family with species occurring all over the world. Mason bees, leaf-cutting bees, and similar species are part of the family Megachilidae. Bees in this family are also solitary and some are cleptoparasites. Megachilids are unique in the way they carry pollen and build their nests. Pollen is carried via structures called scopa on the ventral side of the abdomen. Larval cells in megachild nests are constructed from natural materials: leaves (leaf-cutting bees), soil (mason bees), plant or animal hairs/fibers (carder bees), or plant resin (resin bees). The family Melittidae contains small, usually oligolectic species found only in Africa and limited northern temperate habitats. The smallest family of bees is Stenotritidae. The twenty-one known species in this family are found only in Australia.

All bee species could be considered specialty wasps within the insect order Hymenoptera. Instead of existing as a predator, parasitoid, or scavenger of other small animals, like typical wasps, bees rely on a plant-based protein source: pollen. Protein-rich pollen is collected by bees as food for their larvae. As adults, bees tend to feed on carbohydrates (i.e., nectar) for quick energy. Since pollen is collected by all bees (with the exception of the cleptoparasitic species), bees can be distinguished from wasps by the branched, pollen-collecting, hair-like setae on their bodies along with other structures for pollen transportation.

The value of bees as pollinators arises because their food comes from flowers. When foraging for pollen and nectar, pollen grains inevitably get transported from one flower to another. The only way fertilization occurs in plants is for a pollen grain of a particular species to be successfully transported from an anther (the male flower structure) to a stigma (the female flower structure). If sperm cells carried in pollen successfully enter the egg cells found in an ovary, fertilization occurs and seeds are produced. Obviously, by flying from flower to flower, bees help facilitate the movement of pollen necessary for plant reproduction.

Not all plants are insect pollinated; pollination can occur via wind, water, or other animals. In addition, some plant species are self-pollinated, meaning pollen moves from anther to ovary within the same individual, without the aid of a biotic or abiotic pollinator. Most tropical trees, many temperate shrubs, a large number of desert plants, and a variety of wildflowers rely on bees for pollination. Thus, native bee diversity is greatest in tropical and xeric regions or in habitats with a variety of shrubs and wildflowers. Agricultural plants need to be pollinated too and, while many important crops (e.g., corn, barley, wheat, rice, oat) are wind pollinated, most fruit and vegetable crops are pollinated by bees.

Not all bees are equal in their pollination efficiency, especially in agricultural settings. Native bees tend to be the most efficient pollinators of native plants, but many agricultural plants are not native to the regions where they are grown or they are grown in such large numbers that native bees cannot effectively pollinate every plant. Moreover, native bee density and distribution can vary year to year based on weather, ecological disturbances, flower availability, or a variety of other factors. In addition, most native bees are solitary so they are found in much smaller numbers than those observed in a honey bee hive. Thus, the value of honey bee pollination services arose.

To understand the value of honey bees in the United States, one could simply visit the Central Valley of California in early February as more than thirty billion honey bees arrive, just in time for the ninety million almond trees to begin blooming (Jabr 2013). The bloom lasts about two weeks, and without the newly arrived honey bees it would be impossible for all the trees to be pollinated. California's Central Valley produces 80 percent of the world's almonds and, with such a short window of time for pollination, honey bees are used to supplement the pollinations services of wind and native insects. Without these services, the global demand of almonds would not be met (Jabr 2013).

The arrival of thirty million honey bees to California is not a natural phenomenon like the epic migration of monarch butterflies. Honey bees come to the Central Valley via semi-truck from commercial apiaries across the United States. Each truck can hold about four hundred hives, each containing thirty–fifty thousand bees. At night, when the bees are most likely to stay inside, each colony is placed by forklift along the fields, usually two hives per acre. During the day, the bees do their work—foraging on the pink and white blossoms, collecting pollen and nectar for the hive while unwittingly providing valuable pollination services to the almond growers.

When California's February almond bloom is over, the colonies hit the road, traveling to apple orchards in Indiana, blueberry patches in Michigan, or cranberry bogs in Wisconsin. Some spend the rest of the year migrating up and down the Eastern Seaboard pollinating fruit crops such as cherries, citrus, and pumpkins from Maine to Florida. Some colonies get a chance to "rest" in the summer months, arriving in the Dakotas or Montana to gorge on alfalfa, canola, or clover. This time of plenty results in large stores of honey that gets harvested by the beekeepers and sold in stores across the country. Commercial hives tend to overwinter in warm states like Texas, Florida, or California (Jabr 2013).

Honey and wax were the original products desired by humans from these valuable pollinators. But, over time the bees themselves, especially selectively bred queens, became a commodity. Eventually it was the pollination services of honey bees that became their most valuable product. As the demand for honey bee products evolved, so did the life of a beekeeper. Once existing solely as sustenance farmers, beekeeper life evolved into one of migration. It is a lifestyle where as many as six months each year is spent "following the bloom" to meet the pollination demands of growers in different parts of the country at specific times throughout the year. It is no longer the honey or the wax that is the greatest commodity sold by beekeepers, many commercial beekeepers make over half of their annual income by renting their hives' pollination capacity to farmers nationwide.

European honey bees (*Apis mellifera*) were first brought to North America by colonists in the 1600s. Since hives were kept in vessels made of pottery, straw skeps, or enclosed wooden boxes, early beekeepers extracted honey and wax by killing the colony at the end of the summer (Oertel 1980). This was not an efficient method for large-scale productivity, thus the focus of beekeeping was to supply personal, family, or neighborhood needs, or to collect a product to be used in local trade.

Scientific innovation jumpstarted modern beekeeping in the mid-1800s with the invention of a beehive with moveable frames (Harbo and Rinderer 1980). In the twentieth century, the dimensions of bee hives standardized as beekeepers realized the impact of space within the hive on the overall health of the colony and the amount of honey produced. Soon afterward came the inventions of the wax comb foundation, centrifugal honey extractor, bee smoker, and protective veil. Collectively, these inventions made it possible to keep bees on a much larger scale and the commercialization of beekeeping began (Oertel 1980).

As beekeeping became more commercialized, a market opened for the production and sale of queen bees. Early entrepreneurs sold small boxes that contained eggs and larvae which developed into queen cells when introduced to a queen-less hive (Oertel 1980). Scientists and beekeepers understood the reproductive biology of honey bees as early as the mid-1800s (Harbo and Rinderer 1980), but due to its complexity, beekeepers were unable to breed bees for desired varieties in the way that farmers and ranchers could breed other types of domesticated animals (e.g., horses, cattle). Major obstacles included the fact that insects and their reproductive structures are microscopic, queens mate in flight, an individual queen may mate with multiple drones, sperm is stored for years inside a fertile queen in structures called spermathecae, and queen honey bees do not readily accept mates presented by humans (Watson 1928). For many years, attempts were made to control honey bee mating by either isolating a queen with a particular drone or through forceful insemination. None of these attempts were successful, practical, or replicable on the scale necessary for commercial breeders until specific equipment and a step-by-step process were documented in 1928 (Watson). This scientific milestone paved the way for selectively bred queens to be reared and sold to commercial beekeepers. Selective breeding led to genetic studies and eventually scientific advancement made it possible to sequence the honey bee genome in 2006. In the twenty-first century, genetic research is at the forefront of attempts to understand important honey bee traits including longevity, resistance to disease, and productivity. This research informs honey bee breeding decisions and management techniques which are foundational supports of the multi-million dollar commercial beekeeping industry.

It was not just the development of new equipment that led to the commercialization of beekeeping in the twentieth century. Commercial beekeeping evolved, like many other modern economic practices, out of the Industrial Revolution. One of the most profound effects of the Industrial Revolution was a dramatic increase in human population which can be tied to medical advances and higher standards of living. In 1750, the global population was approximately 760 million people. By 1800 it was 1 billion and by 1900 it was 1.6 billion. The twentieth century saw explosive growth; by the year 2000 the global population was 6.1 billion. The 7-billion-person mark was surpassed in 2011 and projections place the world population at more than 9 billion by 2050. With the world's population growing at such an incredible rate, the greatest challenge will be producing or maintaining the basic resources needed for survival (i.e., food, clean water, space).

Foley et al. (2005) discuss the dilemma of land use in the modern era. With a global population exceeding seven billion people, humans need to balance land uses necessary for humanity (e.g., growing food, gathering resources for shelter, collecting freshwater) with land-use practices that degrade the land (and thus the resources) on which we depend (Foley et al. 2005). One of the most striking changes in land use that has occurred since the spread of the Industrial Revolution is the migration of human populations from rural to urban environments. By the year 2000, half of the world's population lived in cities (Reba et al. 2016). This shift in human habitation patterns has greatly affected natural environments (land, water, and air) and has drastically changed consumption patterns, leading to changes in how and from where food, water, and power are obtained.

Throughout modern history, farming has been essential in sustaining human populations. Before 1900, most agriculture occurred on small-scale sustenance farms, but the industrialization and urbanization that occurred at the turn of the nineteenth century transformed those agricultural norms. The growing human population, centered in cities, needed to rely on fewer farms for food supplies. Data compiled by the USDA National Agriculture Library illustrate the drastic change in the structure and function of United States farms. From 1860 to 2010, the total number of farms in the United States only increased by 7.68 percent. However, total acreage increased by 126 percent, average acreage per farm increased by 110 percent, while the labor force in agriculture decreased by 86 percent and the percentage of the United States population in agriculture decreased from 48 percent to 0.68 percent (USDA 2012). What does this look like in terms of ecosystems? Natural ecosystems have been transformed as meadows and forests were converted to cities, suburbs, and vast commercial farms. Fields of such magnitude cannot rely on the pollination services of native bees alone, despite the fact that native bees (like bumble bees, mason bees, or sweat bees) are often much more efficient pollinators than non-native honey bees. The problem lies in the numbers—incredibly high numbers of bees are needed to pollinate the immense fields of today's farms. Honey bees are the only species whose colonies can persist from year to year in such large numbers and adapt to the changing conditions that result from modern, migratory beekeeping.

This reliability of pollination services by honey bees has actually led to one of the greatest concerns in the modern age of agriculture—the decline of pollinator populations. During the winter of 2006–7, American beekeepers began reporting sudden, inexplicable loss of honey bee colonies (in some cases, beekeepers lost up to 90 percent of their hives). This phenomenon, which came to be known as colony collapse disorder (CCD), was widespread across the United States and persisted in subsequent years. Although some have blamed the decline in pollinator populations on pesticides widely used on large-scale commercial farms, further research has shown that the blame for the decline in honey bee populations is much more complex. It is chronic exposure to multiple stressors which may include poor nutrition (from foraging in monoculture fields), pathogens (including the infamous Varroa mite), pesticides (both topical and systemic), and other environmental stressors (Goulson et al. 2016). It is not unreasonable to assume that Americans' dependence on migratory honey bee colonies (i.e., bees living on truck beds and needing to adapt to one habitat after another throughout the year) is one of those stressors.

Ironically, the root cause of pollinator declines and its solution are one and the same. It was human innovation and advances in technology which led to the Industrial Revolution, urbanization, and unprecedented human population growth. Advances in technology made it possible for large, commercial farms to be developed to meet the food demands of that growing population. But in doing so, these farms became dependent on the pollination services of just one bee species which led to widespread decline of that one species. Human populations will continue to grow. Demand for food will increase. Modern agricultural practices will become insufficient. And new challenges will arise including impacts of climate change and fluctuations in governmental policies, priorities, and funding. It will have to be human innovation and advances in technology that solve these problems for future generations.

The innovative problem solving has already begun. Entomology continues to be an important field of scientific research. DNA technology is being used to understand variations and adaptations within honey bees and to use that understanding to selectively breed queens that are resistant to pathogens and environmental stressors. Ecologists are looking at the benefits of breaking up large monoculture fields with native flower strips to supplement the diet of honey bees and to create habitat and forage for native pollinators. Entomologists are elucidating the pollination services of other bee species and looking at ways to produce them for use on commercial farms. Some, like the blue orchard mason bee (*Osmia lignaria*) and the common eastern bumble bee (*Bombus impatiens*) are already being produced and sold for use in greenhouses and agricultural fields.

At the same time, pollinators, especially bees, have become elevated in the eyes of the public to the status of cultural icon. "Save the Bees!" is a slogan commonly found on T-shirts, coffee mugs, protest signs, billboards, and even cereal boxes. But the save-the-bees movement is not solely about bees. Starting in the 1960s, with the publication of Rachel Carson's book, *Silent Spring*, public awareness grew along with concerns about air, water, land, and noise pollution, the use of certain pesticides, waste disposal practices, extraction of natural resources, and a host of other environmental issues. Since then, various animals have been used by conservation organizations as environmental ambassadors. In the 1970s to 1980s, cries of "Save the Whales!" were popular as both the impacts of the commercial whaling industry and the delicate balance of the ocean ecosystem were brought to the public eye. Other organizations have used tigers, polar bears, or toucans to raise both awareness of and funds for conservation projects across the globe.

The biology of bees is fascinating and their ecological value is substantial, so their story has been an easy one for scientists to tell and for citizens to get excited to share. Thus, when honey bees started inexplicably dying and native pollinator populations were shown to be in decline, the world took notice. Since the early 2000s, news and social media have propagated these insects into the role of poster child for all things environmental and sustainable. Protestors dressed in bee costumes have "swarmed" to demonstration sites, promoting pesticide bans, opposing certain urban development projects, advocating for environmental regulations, or highlighting negative impacts of pesticides. Bees are used in advertisements and cartoons to promote sustainable agricultural practices. In 2017, General Mills even pulled "Buzz," the famous Honey Nut Cheerios mascot, off cereal boxes in an effort to highlight concerns over declining pollinator populations (Mills 2017). The plight of the bees has been used to promote a plethora of products from specialty honey and beeswax candles to bumper stickers, T-shirts, coffee mugs, wildflower seeds, and amateur beekeeping gear. In the midst of the save-the-bees movement, hobbyists, conservationists, and educators are working hard to raise both money and awareness for the importance of insect pollination (by both honey bees and native, wild bees) while some government and industry leaders are trying to create a cultural shift from consumerism to sustainability (Worldwatch Institute 2010).

CONCLUSION

Insects, the physical products they produce, the innovation they inspire, and the services they provide are incredibly valuable in the modern era. As technology and scientific understanding continue to progress into the future, there is no doubt that the value humans place on insect products will continue to increase. Although current population growth, consumption patterns and land use can certainly have negative effects on ecosystems and biodiversity, the situation is not so dire that we should lose hope. Research, innovation, and public education are the keys to understanding the scope and diversity of the insect world, realizing the value of insects and their products, and innovating ways for humans to use insect products in a sustainable way.

CHAPTER FIVE

Insects in Mythology and Religion

GARY NOEL ROSS

Artifacts and historic writings bear testimony that insects have played major roles in cultures throughout the ages and throughout much of the world (Beutelspracher 1988; Rothschild 1991; Ross 1997a; Manos-Jones 2000; Nazari 2014; Nazari and Evans 2015). But whether insects are part of age-old mythology and religion, or simply cursory artistic expressions, is often unclear.

For many Native Americans, however, there is no ambiguity. Educated as a field entomologist and ecologist, I have been fortunate to interact with thirty-six Indigenous cultures throughout the American Southwest, Mexico, and Central America. My introduction traces back to June 1962. As a nascent graduate student at Louisiana State University (USA), my research centered on an ecological study of the butterflies of the Sierra de los Tuxtlas, a little-known volcanic range in southeastern Veracruz, Mexico. During one of my field trips to a particularly isolated region, I happened upon an American missionary family living in a small village of Indigenous people known as the Sierra Popoluca. Following that auspicious encounter, I spent a total of fifteen months between 1962 and 1967 as guest within the world of the Popoluca.

My exposure to the Popoluca during my formative years triggered a life-long fervor for Indigenous peoples—especially their handicrafts and their relationships with butterflies. To accommodate this resolve, I have periodically sought out other Native American cultures. And although modern technology has altered the mores of even the remotest of peoples, I have learned that one animal continues to be singled out as the irrefutable sovereign of mythology and spirituality: THE BUTTERFLY.

That an insect from the Order Lepidoptera (butterflies, moths, and microlepidoptera) would become such a luminary is not perchance. Consider, the butterfly's beauty and graceful flight have earned it the endearing descriptor "charismatic megafauna of the insect world" (Schappert 2000: 12) and "the most beautiful creatures in all of creation" (Ross 1995: 42). Then there is the butterfly's resemblance to the benevolent angels and fairies of religion, literature, and art. Finally, the metamorphic life cycle that transforms ovum (egg), to larva (caterpillar), to pupa (chrysalis), and finally to adult (butterfly) seems magical (Feltwell 1986: 17–29; Ross 1995: 23–6; Schappert 2000: 22–40). The overt differences between each stage have intrigued *Homo sapiens* for millennia. Gene Kritsky and Ron Cherry (2000: 30, 35, 36), for example, report that of the 222 myths studied involving metamorphosis, the Lepidoptera constituted the number one taxon (37 percent) and their role in metamorphosis was the most geographically widespread (six continents); Coleoptera (beetles), Diptera (flies/gnats/midges), and Hymenoptera (bees/wasps/ants) followed, respectively.

With such panoply and allure, the butterfly was destined to acquire a sacred persona. Dame Miriam (1908–2005), the august English naturalist and author who devoted a lifetime of study to the butterfly-human liaison, concluded that as far back as ancient Greece (800–500 BCE), the butterfly and the human psyche have been inextricable (1991: 7–12). In fact, the English word "psyche" traces directly from the ancient Greek *psyche*, a word that denotes both the human soul (an individual's essence of life) and the insect we note as "butterfly." In Greek mythology, the Goddess *Psyche* was the wife of *Eros*, God of Love who was always illustrated as a butterfly. In the words of Rothschild: "… in many lands and at different times the angelic butterfly was a symbol of the soul" (Rothschild 1991: 12). Such sacredness evolved into the cornerstone for what Rothschild came to describe as "butterfly worship" (52). More recently, Vazrick Nazari (2014: 228) reinforced the idea that for untold millennia, butterflies have been the quintessential metaphor for transcendence—that is, the transformation of the human spirit after corporal death to a higher existence. In modern techno-savvy societies, however, that tradition has become but a specter of its former self. Be that as it may, day-to-day practices of native peoples are a window into ancient libraries, which in turn, resonate into modern lifestyles.

What follows are six treatises describing my personal experiences with the butterfly-human relationship in six contemporary Native American cultures. These are: Sierra Popoluca culture of Veracruz, Mexico; Mazahua, Otomi, and Tarascan cultures of Michoacán, Mexico; Zapotec culture of Oaxaca, Mexico; Maya culture of Southern Mexico and Guatemala; Miscellaneous Indigenous cultures of Mexico; and Zuni culture of New Mexico, United States of America. These essays are not meant to be a universal/definitive treatise on the "Cultural History of Insects in the Modern Age." Furthermore, because a mindfulness of the past helps to understand the present, I have included relevant historic references.

PERSONAL EXPERIENCES

Sierra Popoluca Culture of Ocotal Chico, Veracruz, Mexico

The Sierra Popoluca ("Popoluca" for short), or *Nuntajɨypáppɨc* ("straight-speaking-ones"), which I first encountered in 1962 (Ross 1975–7, 2010c), identified their homeland to be the Sierra de Los Tuxtlas (Tuxtlas Mountains) in Mexico's Gulf coast state of Veracruz (Soriano et al. 1997). These Native Americans lived in villages on or near the dormant Volcán Santa Marta, elevation 5,577 feet (1,700 meters) and Volcán San Martín Pajapan, elevation 3,806 feet (1,160 meters). The epicenter of native speakers was Ocotal Chico ("Little Piney Ridge") located at 1,800 feet (549 meters). An American family, John and Royce Lind—members of the Summer Institute of Linguistics (now SIL, International) and the Wycliffe Bible Translators (SIL 2017)—had taken up residence in Ocotal Chico in 1960 to study the unwritten Popoluca language and to translate the New Testament of the Bible into the Popoluca idiom.

In the 1960s, the Popoluca consisted of but a few thousand individuals living in approximately thirty small settlements (Ross 2010a, 2010b, 2010c). This was an oral society: no writing, no hieroglyphs. In addition, artistic expressions were limited: 1) an undecorated basket woven by an elderly gentleman for use in draining and washing boiled corn; 2) simple flower-like etchings on half-gourds used for ladling water; and 3) simple geometric etchings on clay pots used to store small household items. Such dilution of the culture was due to the influence of Spanish-speaking communities in nearby lowlands.

Many anthropologists theorize that the Popoluca are the direct descendants of the Olmec (1200–400 BCE)—the seminal and enigmatic culture within Mesoamerica (a geographical region of Mexico and Central America in which cultures flourished before 1519 CE, the pre-Hispanic/pre-Columbian Era. Within the realm of the Popoluca, several Olmec artifacts of basalt have been verified, including a massive anthropomorphic figure in a crouched position (Bernal and Groth 1968: 26–7; Grove 2014: 75–9). Because the Olmecs were a purely archeological culture, artifacts have no recognizable inscriptions. Nevertheless, all were revered to some degree by the Popolucas.

That aside, the indigenes retained a kinetic connection to the supernatural. This included the belief that butterflies (*me̱me*) are magical creatures that harbor good spirits. The common and spectacular blue morpho butterfly

(*Morpho helenor montezuma*) (Lamas 2004: 1955–96), for example, is called *cawaj me̱me* ("horse butterfly") (Elson and Gutiérrez 1999: 12), no doubt referencing the butterfly's low gallop-like flight. The belief is that when a blue morpho enters your house or flies about you, the sighting is an announcement of an imminent visit by someone. By contrast, moths, *me̱mtúchti* ("butterflies of the night") (Elson and Gutiérrez 1999: 83), were omens of evil.

I quickly learned that Popolucas equated spirits to witchcraft, that is, power over good and evil. Because I was collecting butterflies, villagers touted me as a de facto witch doctor (shaman) capable of casting spells. Labeled *me̱me pi̱xiñ* ("butterfly man") (Elson and Gutierrez 1999: 83, 147), children were excited to assist with my collecting, but kept a cautionary distance. Initially, youngsters were proud to bring me specimens they had beaten to the ground with leafy twigs. But wings were missing (young Popolucas often de-wing butterflies before consuming them as bite-size snacks). Aware that my young friends had no concept of science, I politely informed them that I did not eat butterflies. Instead, I collected the insects to show my family and friends back home the beauty of the Popolucan world (Ross 2010c: 121–2). The simple explanation worked. I then instructed my apprentices with a technique for handling butterflies that did not injure their wings. The next day the innovative children crafted small nets from forked twigs and spider webs; to my surprise, the crude nets proved quite serviceable.

Children often tailed me during my night forays carrying a kerosene lantern onto a pine ridge outside the village. My activity centered on research into the life history of an unrecognized myrmecophilous (ant-associated) species of metalmark butterfly (family Riodinidae) (Clench 1964; Ross 1964a, 1964b, 1975–7, 1985, 1994, 1997b, 2010a, 2010b; Soriano et al. 1997: 323–8). I was surprised that neither children nor adults had any knowledge of this small-scale "animal husbandry" on the outskirts of their village. But by assisting the *me̱me pi̱xiñ*, my young assistants gained standing among their playmates and their parents learned a bit more about their homeland.

The local shaman, however, was not so enamored. This spiritual sage would go out of his way to avoid a face-to-face encounter. My American hosts surmised that this behavior stemmed from envy and fear because he believed me to be a rogue "man of the spirits" (Ross 2010c: 121).

My perceived spiritual prowess was heightened after my first extended excursion to the crater of Volcán Santa Marta (Ross 2010c: 121). For Popolucas, the pristine forest cloaking the upper slopes of the volcano harbors malevolent night spirits. The belief became evident when I and my field partner, Mike Lind, the young teenage son of my host family, first departed Ocotal Chico for a four-day camping saga to the crater. Locals were quick to regale the Linds: "You will never see them again." When Mike and I returned, villagers lined the trail to cheer. In Popoluca context, if I could survive the demons of the forest, I patently was connected to the spirit world. From then on, I was a celebrity.

Today, the Sierra Popoluca population has increased to roughly thirty thousand distributed within fifty-eight communities (personal communication). From an environmental perspective, however, what was is no more (Dirzo and Garcia 1992; Soriano et al. 1997: 60–85; Durand and Lazos 2008; Ross 2010c: 124–5). Researchers estimate that 85–90 percent of the original forest on the Tuxtla volcanoes has been removed. But that has slowed since the 1998 federal government's establishment of *Reserva de la Biosfera Los Tuxtlas* ("Los Tuxtlas Biosphere Reserve"). This protected land is respected by the Popolucas. Local men, in fact, often assist federal officials with the replanting of native trees.

And the future for residents of Ocotal Chico? Globalization has exposed the erstwhile provincial Popoluca to amenities of the twenty-first century (Ross 2010c: 124–5). Emboldened by salient literacy (the result of the long-time commitment by the Lind family coupled with incentives from the Mexican government), the modern *Nuntajɨypáppic* now enjoy a bulwark against extinction. How butterflies will feature in this revitalization, of course, is unknown.

Mazahua, Otomi, and Tarascan Cultures of Michoacán and Mexico, Mexico

Throughout much of Mexico, *Día de los Muertos* ("Day of the Dead") is the juggernaut of fiestas. The event, which has its origins in Aztec mythology, is annual and celebrates the temporary return of ancestral spirits. In ancient mythology, *Mictecacihuatl* ("Goddess/Lady/Queen of the Dead"), ruled over death (Maza 1987: 39–44; Beutelspracher 1988; Manos-Jones 2000: 16–22). The deity guarded the bones of the dead and presided over celebrations for the dead. In time, such pagan traditions were commingled with spirit-oriented ideology and canon introduced by Spanish Roman Catholic missionaries. Currently, *Día de los Muertos* is a three-day pan-religious celebration stretching over "All Hallows Eve" (October 31), "All Saints Day" (November 1), and "All Souls Day" (November 2).

To prepare, in early October the pious tidy up cemeteries, churches, and home altars. The bright yellow and aromatic flowers of native marigolds (*Tagetes lucida*)—symbols of death—are often placed in sacred places as part of pre-fiesta protocol. Items of food and drink are also placed at most sites for the pleasure of the departed and as an entreaty to grant favors to the living. By October 31, cemeteries are awash with a collage of bright colors as well as fresh fruits, sweets, and beverages. Vendors hawk candles, paper cut-outs, streamers, and papier mâché caricatures. Candies and sugar-coated breads decorated with macabre motifs of skulls and coffins are common.

Festivities peak on November 2. Schools and businesses close, parades are launched. Locals dress in costumes depicting skeletons. Fireworks bombard the carnival-like aura to drive off evil spirits. The next day offerings are cleared. Because perishable items dehydrated and are free of aromas, locals believe that the visiting spirits partook of the essences of the gifts.

Danaus plexippus: monarch butterfly (Urquhart 1987; Malcolm and Zalucki 1993; Schappert 2004). Medium-size, wings orange and black flecked with white, and a flight pattern that is slow and soaring, the monarch is truly "regal" to behold. And yet, it is the species' annual marathon migrations that have been lauded as "the most spectacular in the insect world" (Malcolm 1987: 135; Vane-Wright 1993: 179) and tagged as "a unique biological phenomenon" (Brower 1977: 52; 1985, 2015: 29; Vane-Wright 1993: 179) and an "endangered phenomenon" (Brower and Malcolm 1989). It is no surprise that the monarch has become the charismatic poster child for the international conservation movement (Halpern 1998; Brower 2001; Pyle 2001: 25; Sauer 2001; Schappert 2004: 28).

During each September and October, tens to hundreds of millions of these flashy butterflies from the eastern United States and eastern Canada embark on an improbable flight southward to Mexico where they spend the cooler months of winter. During this southern migration, individuals have been recorded to fly a distance of at least 2,400 miles (3,862 kilometers) (Brower 1977: 52; Ross 1995: 29; Brower 2015: 28). The following March–April, these international travelers begin winging their way northward, laying eggs en route on milkweed plants (*Asclepias* spp.)—the monarch's sole plant capable of nourishing its larvae (caterpillars). The butterflies then die, relying on subsequent generations to migrate southward the subsequent fall. (No individual survives an entire year to repeat a migration.)

But the winter bastion for the eastern monarchs remained clandestine to scientists. That mystery was solved on January 9, 1975. An American-Mexican couple (Ken and Cathy Brugger), who were laboring under the tutelage of Canadian monarch scientists, Fred and Norah Urquhart, were scouting in the massive highland known as the Transverse Neovolcanic Belt that straddles the interior states of Mexico and Michoacán. In the immortalized words of Ken Brugger, there were "millions of Monarchs in evergreens beside a mountain clearing" (Urquhart and Lavies 1976: 166; Urquhart 1987: 153).

The smoking gun is the ancestral home of the hunter-gatherer Mazahua, Otomi, and Tarascan peoples (for reviews, see Urquhart and Lavies 1976; Urquhart 1987: 153–6; Ross 2001: 30–2; Schappert 2004: 21–32). Remoteness, high elevations, and cool temperatures kept outsiders at bay. But the January 9 discovery quickly made international news. Featured as the cover article of the August 1976 issue of *National Geographic* magazine (Urquhart and Lavies 1976), the migratory monarchs were propelled into the limelight.

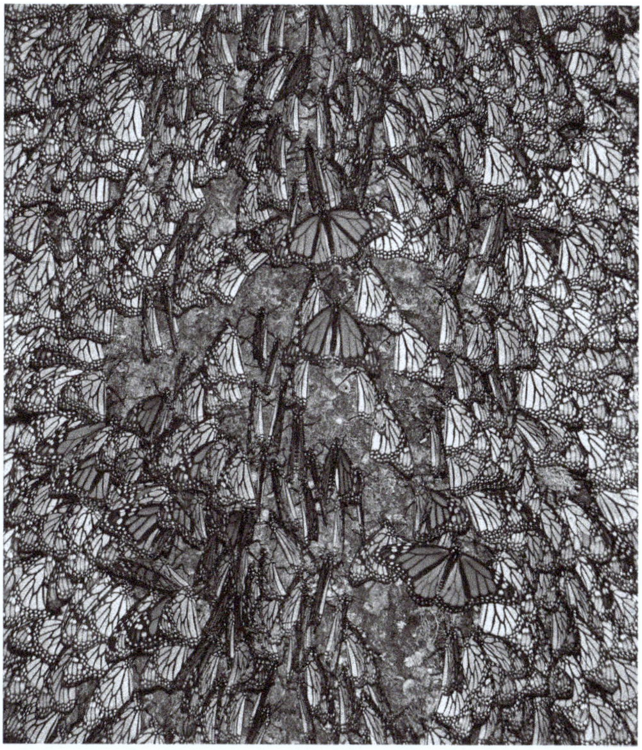

FIGURE 5.1 Section of trunk of an oyamel fir (*Abies religiosa*) tree with a cluster of overwintering monarch butterflies (*Danaus plexippus*). "Monarch Butterfly Biosphere Reserve," Michoacán, Mexico. Photograph courtesy of Gary Noel Ross.

Today we understand that the high-altitude fir forests of the states of Mexico and Michoacán are the consistent winter haven for thirty-five to six hundred million monarch butterflies (Williams and Brower 2016).

The storied winter monarchs languish in a semi-torpid state from late October to mid-March. The butterflies are concentrated in a dozen or so forest enclaves at cool altitudes between ten thousand and twelve thousand feet (3,048–3,658 meters). There, the ectothermic insects are wedged onto the branches, trunks, and needles of trees—giving the impression of autumn leaves or perhaps, that butterflies are consuming the trees. The targeted tree is the cold-tolerant, evergreen, and limited oyamel fir (*Abies religiosa*). The conifer's dense/short needles and rough bark provide tenacious footholds for the butterflies' tiny, clawed feet. But when the sun shines, the butterflies take flight en masse for exercise and to secure water. The aerial spectacle has been compared to "orange snow" (Schappert 2004: 22). Additionally, the beating of myriad wings produces a soft "chatter" and a "rain" of fairy-tale pixie dust—the latter due to dislodged wing scales (Ross 2001: 32).

When monarchs first settled into the high-mountain forests is, of course, an unknown. Scientists theorize that the monarch migration as we know it could not have existed earlier than ten thousand years ago; before that time much of North America was gripped in glacial ice, and that patently precluded milkweed and monarch habitats (Vane-Wright 1993: 180).

Today, in nearby communities, the monarch butterfly is the reigning celebrity during the annual *Día de los Muertos* festivals (Ross 2001: 30; Schappert 2004: 85). Activities include virtually all hallmark celebrations elsewhere except that the costumes featuring skeletons are adorned with bright orange/black wings—the signature colors of monarchs. Additionally, toys, banners, handwoven Mazahua/Otomi textiles, and food items are colored or accented with these bona fide hues.

But while locals presumably were intimate with the annual arrival and departure of the monarchs, the butterfly's complex biological cycle remained a mystery. Local narratives focused on: 1) butterflies materialize in a secreted locality each year; 2) butterflies always arrive on or near the national *Día de los Muertos*; 3) butterflies cluster on trees with pyramidal shapes that are reminiscent of the spires atop towering cathedrals; 4) the tips of the trees' branches are configured in the form of a Christian cross; 5) the trees produce a resin that when burned releases an aroma similar to the incense used in the Catholic church; 6) butterfly clusters are analogous to congregations of worshipers in a Christian church; 7) food offerings lose their essence—proof that the gifts are accepted; 8) each spring, the transient insects depart—presumably to return to their spiritual realm. In summary, the drama validates the transcendental nature of the human spirit, and honors an anachronistic tradition that connects the living monarch butterfly to mythology and spirituality that predate history.

The government of Mexico was quick to herald the monarch colonies as a "living national treasure." In 1980, eight of the fourteen colonies were cordoned off into a protected zone designated *Reserva de la Biósfera de la Mariposa Monarca* ("Monarch Butterfly Biosphere Reserve"). The colonies cover 11.6 acres (4.7 hectares) of the total reserve of 139,019 acres (56,259 hectares). Five colonies are open to visitors: Sierra Chincua and El Rosario in the state of Michoacán, and La Mesa, Piedra Herrada (Valle de Bravo), and El Capulin in the state of Mexico. El Rosario is the largest sanctuary; there the butterflies cover approximately 1,500 trees. In 2008, the sanctuaries were designated a "UNESCO World Heritage Site" (Monarch Biosphere Reserve 2017). Each winter the international reserve hosts tens of thousands of tourists, principally from Canada, France, Germany, Japan, Spain, and the United States. In addition, since 1992 the colonial mining town of Angangueo celebrates each February the "Festival de la Mariposa Monarca" (Monarch Butterfly Festival). The event promotes the importance of butterfly habitat, the culture and arts of the area, and the importance of ecotourism to local economies. Workshops include discussions of the mythology and spirituality of the monarch. As lagniappe, attendees can purchase memorabilia, especially papier mâché figures of human skeletons with "Monarchesque" wings.

What of the future? The modern world has intruded into the monarch microcosm. Each winter, ecotours from Canada and the United States visit two sites within the sanctuary that have become popular: El Rosario and Sierra Chincua. Throngs of tourists bring in badly needed finances to the region. In addition, natives are being educated to the butterfly's life cycle and the importance of maintaining the oyamel forest. For the butterflies, however, the continued removal of trees and a warming climate are causing calamitous vicissitudes (Schappert 2004: 28–32; Butler 2015; Oberhauser et al. 2015). Thus, the futures of both native peoples and monarch butterflies are uncertain. But for now, international tourists flock to the region, and natives continue to honor their ancestral traditions.

Having visited the butterfly colonies on four occasions, I can attest that such a tableau evokes strong emotional responses from tourists. Adults stand transfixed in silence. Breathing slows. Eyes glaze over. Children become giddy and point their fingers ceaselessly. "Awesome," "breathtaking," "divine," "enchanting," "entrancing," "glorious," "hallowed," "magical," "mesmerizing," "riveting," "sacred," "spellbinding," "spiritual," "sublime," "surreal," and "wondrous" are common exclamations. The suite of rarefied/misty air, majestic dark verdant forest, millions of butterflies, soft sounds of fluttering wings, and sprinkling of tiny golden wing scales all superimpose themselves to elicit a cascade of visceral sensations so that you don't have to be a particularly religious person to be inspired. As such, I deem this adoration of the monarch butterfly as the prime example of modern butterfly worship on an international scale.

Turns out Mexico is not the only venue where monarch butterflies have undergone what I refer to as "monarch sanctification." West of the Rocky Mountains, autumn monarchs migrate westward, congregating in at least 253 monitored Pacific coast woodlands stretching from Mendocino County in northern California to Ensenada, Baja California, Mexico (Lane 1993: 335–44; Schappert 2004: 17–19; Sarikonda 2016: 7). There, monarchs prefer non-native eucalyptus trees (family Myrtaceae) as well as native Monterey pine (*Pinus radiata*) and Monterey cypress (*Cupressus macrocarpa*). In 1997, the western monarch population was assessed at just over one million. But by 2016, the number had dropped to under three hundred thousand (less than 1 percent of the lowest number for eastern monarchs). Pacific Gove, a coastal community in Monterey County, California, is nicknamed "Butterfly Town U.S.A." (City of Pacific Grove 2017). There, the "Pacific Grove Monarch Grove Sanctuary" (PGMBS) protects the butterflies. On the first Saturday in October, costumed children parade to welcome the butterfly visitors. The celebration, which dates back to 1939, is accompanied by a Butterfly Bazaar. The Pacific Grove

Museum of Natural History offers programs on monarch biology, mythology, and spiritual symbolism. According to many visitors, their experiences trigger a serendipitous "spiritual awakening."

Zapotec Culture of Teotitlán Del Valle, Oaxaca, Mexico

The epicenter for Zapotec culture is the semi-arid mile-high plateau in the southern state of Oaxaca, Mexico. Zapotecs are extremely proud of their ancestry that predates the more familiar Aztec culture (Whitecotton 1977). As far back as 500 BCE, Zapotecs were employing the portable backstrap or "stick" loom to weave quality textiles: yarns from native plant fibers, silk from a local social caterpillar (Fitzgerald 2001), and dyes with natural pigments. The industry was Mexico's equivalent of Parisian *haute couture.* Because of this distinction, the Zapotecs were inured by the Aztecs as an elite tribute state. (For reviews, see Ross 1986, 1987, 1988, 1997a, 2012a, 2012b, 2015a; Stanton 1999; Manos-Jones 2000; Selcraig and Borris 2003.)

April 21, 1519. Two worlds collide. Spanish Conquistadores and Christian missionaries disembark onto the east coast of Mexico. Driven by zeal, pragmatism, and a "Colonial Imperative" rather than righteousness, the Spanish looked upon everything in the exotic land with a pejorative mindset. Their prime directive was abject intrusiveness, that is, replace all native spiritual practices with Roman Catholic doctrines. The result? The lives of native peoples in the New World were altered forever.

Then again, the alien iconoclasts were not all bad. The Spanish, for example, introduced churra(o) sheep (*Ovis aries*) and the fixed frame (upright) pedal loom. (Wool is more durable than cotton and provides an expanded palette of color, and the pedal loom allows for the weaving of larger textiles.) These innovations revolutionized textile production. One of the "new items" was a utilitarian textile. Labeled *serape* (blanket), the garment became a convenient body wrap and bed cover. With the discovery of synthetic (chemical) dyes in Germany in 1856, a plethora of novel colors further augmented the hues of natural wool and natural dyestuffs.

The village of Teotitlán del Valle ("Valley Home of the Gods") or *Xaguixe* ("At the foot of the Mountains") boasts about six thousand inhabitants. Almost all flaunt pedigrees of pure Zapotec, and almost all are imbued with acumen for weaving. Between 1976 and 1982, I headquartered for ten months in Mitla, Oaxaca. During that period, I often visited nearby Teotitlán, focusing my interactions with Isaac Vásquez Garcia (b. 1935–2022) and Alberto Vásquez Jiménez (1941–2007), first cousins and celebrated masterweavers.

"As a youngster I was curious about the past," Isaac narrated. "By questioning village elders, I gleaned recipes for six native dyes." In time, the astute Isaac navigated through museums, books, and archeological sites to familiarize himself with ancient designs. Isaac's research scoured Aztec (Nahua), Maya, Mixtec, Teotihuacán, Toltec, and Zapotec cultures dating as far back as 100 BCE to 700 CE.

Isaac likened the images to muses awakening. With such milieu, Isaac and his younger cousin, Alberto, were soon creating textiles featuring facsimiles of gods, animals (especially butterflies), plants, and geometric designs from yesteryear. Each piece demonstrated virtuoso craftsmanship: virgin wool, handspun yarns, tight construction, and refulgent colors from natural dyestuffs (for reviews see Ross 1986, 1987, 1988, 2012a, 2012b, 2015a, 2015c; Stanton 1999; Selcraig and Borris 2003).

The innovative textiles quickly achieved stardom as a new modality: the *tapete* (tapestry or rug). And there was more. Besides ancient motifs, several aspiring artisans introduced images from other cultures: Navajo (from the American Southwest) and European modern. Such stellar weavings are now considered objets d'art. They are viewed in situ by as many as a half-million tourists a year, featured in international videos and printings, collected by intergenerational internationals, and held within the collections of major museums (Stanton 1999: 2; Selcraig and Borris 2003: 74). What was once a mom and pop cottage

FIGURE 5.2 Tapestry of a mythical butterfly from a pre-Hispanic clay stamp excavated at Tula, Hidalgo, Mexico. Toltec/Aztec culture. Virgin wool and natural indigo dye. Woven on a fixed frame (upright) pedal loom by masterweaver Isaac Vásquez Garcia. Photograph courtesy of Gary Noel Ross.

business focused on utilitarian textiles and tourist souvenirs was transformed into a multimillion-dollar art industry involving hundreds of families.

Today, Isaac Vásquez's salon flaunts the English name "THE BUG IN THE RUG." The name is a reference to the cochineal insect dye featured in my article of that same name published in *Natural History* magazine in 1986. A step into Isaac's gallery is a step into a showcase of kaleidoscopic phantasmagoria where tradition and modernity come together. Of course, the butterfly image reigns supreme.

FIGURE 5.3 Tapestry of Mayan Goddess painted on a pre-Hispanic ceramic excavated at Palenque, Chiapas, Mexico. Virgin wool and natural dyes. Woven on a fixed frame (upright) pedal loom by masterweaver Isaac Vásquez Garcia. Photograph courtesy of Gary Noel Ross.

Although present-day Zapotecs do not acknowledge butterfly worship *sensu stricto*, Sr. Vásquez eloquently voiced:

> We Indians don't know much about our past. The images as recorded in the artwork of the past reflect the beliefs and customs of those people at those times. That is why I prefer to weave ancient images. By weaving tapestries as fine as I can with ancient designs, I hope that current and future generations of my countrymen will find a new connection to and an appreciation, reverence and respect for their cultural heritage. Maybe, also, foreigners will come to understand what we Indians have contributed – and continue to contribute – to the world.

To understand the Zapotec preoccupation with butterfly motifs—what I like to call "nostalgic timelessness" and "memories in memoriam"—one must backtrack to pre-Columbian Mexico. Then, the central plateau was dominated by Teotihuacán culture (100 BCE–550 CE), Toltec culture (900–1168), and later, Aztec culture (1240–1519). All honored a pantheon of deities (polytheism), and all relied on psychoactive plants to enhance their sacred rituals.

Archeologists know little about the Teotihuacános and Toltecs. But because Aztec civilization was at its apogee, much of Aztec life is legend (Whitecotton 1977). Information has surfaced from extant codices (sacred folded-screen picture books chronicling history—real and mythological), painted murals, carvings on stone, stone monoliths (stelae), architectural glyphs, friezes and frets, ceramics (especially funerary urns), feather-work, personal adornments, and clay stamps (used for embossing ceramics, textiles, and tree bark/deer hide paper) (Beutelspracher 1988; Ross 1997a, 2012a, 2012b).

When it comes to divination, Aztecs embraced a collective consciousness that was butterfly-centric (Beutelspracher 1988; Manos-Jones 2000: 16–22). The butterfly was called *papáotl*, butterfly eggs, *ahuauhpapálotl* ("seeds of happiness"), and caterpillars, *ocuilpapálotl* ("worms of butterflies"). *Xochiquetzal* (literally "precious flower")—sometimes referred to as *Xochiquetzalpapálotl* ("flower of the precious paradise butterfly")—was the Mother Goddess who was represented in the mortal world by the two-tailed swallowtail butterfly (*Papilio multicaudata*). The Goddess reigned over aesthetics, love, flowers/vegetation, domestic laborers, fire, and spirits of the dead. She accompanied warriors into battle, making love to them while holding a butterfly between her lips. (At death, the warrior's spirit was transformed into a butterfly before being escorted into the paradisiacal afterlife.) In addition, *Xochiquetzalpapálotl* was the mother of *Quetzalcóatl* ("feathered serpent"), the God of Life and the Lord of Dawn who gave birth to humankind. In one legend, warriors at death joined with the sun as companion with the reborn *Quetzalcóatl*, and after four years were transformed into hummingbirds or butterflies. A third Goddess,

Itzpapáotl (the "Obsidian Butterfly Goddess"), was illustrated with four knives, and recognized in the real world as the large saturniid moth, *Rothschildia orizaba*. The Goddess represented movement/travel, the hunt, human sacrifice, war, celestial Venus, Earth, and Moon; she also was the patron of women who died in childbirth. Most of these deities were associated with the transformative nature (metamorphosis) of the Lepidoptera.

Deities were often depicted as a God-butterfly hybrid—a mythological avatar. This nexus is theorized to portray the bridge between the ethereal world of the spirits and earth-bound man. In addition, images of single butterflies were rife—particularly in clay stamps (Enciso 1953, 1971; Field 1974), painted murals, and ceramics (Ross 2012a, 2012b). It is this treasure trove of artifacts that continue to inspire Zapotec weavers.

Meanwhile, the *tepetes* from Teotitlán del Valle can be viewed not only as sentimentality on a grand and timeless scale, but as ambassadors for promulgating the art, mythology, and spirituality (including butterfly veneration) of historic Mexico. Another positive upshot is that the Indigenous village is reputed to possess the highest standard of living of any Indigenous community in the Western Hemisphere, if not the world (Selcraig and Borris 2003: 72).

To pay homage to Mexico's heritage of artistry and butterfly worship, I orchestrated an ambitious exhibit of Zapotec weavings in the gallery of the Arts and Humanities Council of Greater Baton Rouge (McMullan 1983; Price 1983; Ross 2012b) between June 8 and July 8, 1983. Titled "The Zapotec Tapestries: Translations of Pre-Columbian Myth," the exhibit showcased sixty-two textiles and several hundred photographs from Teotitlán. During the first two weeks, Sr. Isaac Vásquez G. was present to interact with visitors through workshops and tours. Later in 2004, I arranged a two-year (2004–6) exhibit titled "Zapotec Tapestries" to inaugurate the McGuire center for Lepidoptera and Biodiversity, Florida Museum of Natural History, University of Florida (Gainesville). The display presented thirteen tapestries of the Vásquez cousins (Ross 2009, 2012a, 2012b).

Maya Culture of Yucatan, Chiapas, Mexico and the Highlands of Guatemala

Along with the Aztec, the Maya were a bastion of Mesoamerican culture (Peterson 1976; Morris 1987). Mayan geography included southern Mexico from the Yucatán Peninsular and Chiapas, then south through Guatemala, Honduras, and a much of remaining Central America.

Most Mayan ceremonial centers (250–900 CE) had been abandoned centuries before the Spanish first arrived at the coast of Yucatán in 1511 (Morris 1987: 19). Today, many of these—Bonampak, Calakmul, Caracol, Chichén-Itzá,

Copan, Palenque, Tikal, Tulum, Uxmal, Xunantunich, and Yaxchilán, for examples—are major archeological sites.

Unlike in Aztec culture, images of butterflies are not widespread in Mayan artifacts. Hieroglyphs usually illuminate the chronologies of Maya kings as well as astrological events and calendars. Even so, within the four extant codices, there is a legacy of butterflies: several humanoid figures are accented with butterfly-like apparel.

That said, there are numerous examples of butterfly imagery in contemporary and antique Mayan textiles (Morris 1987: 106; Ross 2015a), especially the *traje* (traditional woman's tunic or *huipil*) (Cordry and Cordry 1968: 50–80). These are dyed with vibrant aniline dyes—especially red imitations of the native cochineal dye (see the previous section on the Zapotecs)—and woven on a pre-Hispanic style backstrap loom (Cordry and Cordry 1968: 25–49). Stylized butterflies are common in weavings of the Tzotzil Maya community of Magdalenas located in the highlands of Chiapas. The insects are associated with

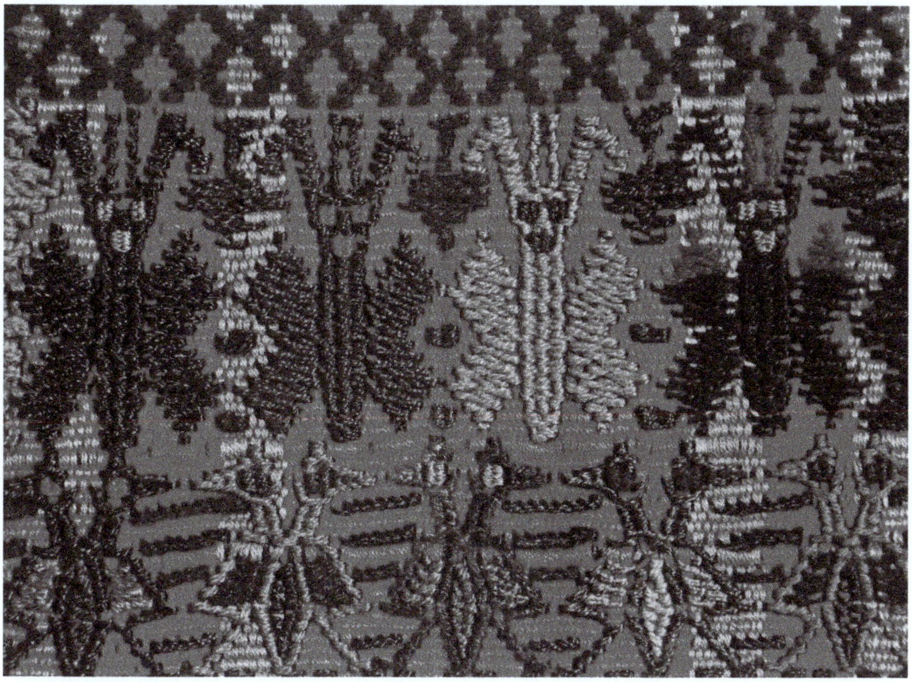

FIGURE 5.4 Contemporary woman's sash from Santa Catarina Palopó, Guatemala. Cotton/synthetic fibers and synthetic dyes. A row of butterflies is bordered by a row of two-headed eagles and a row of diamonds. Woven on a pre-Hispanic backstrap loom by an unidentified local female. Cakchiquel (Maya) culture. Photograph courtesy of Gary Noel Ross.

motifs representing the sun (representation of good, and the antithesis of the underworld); designs are usually bold—even a bit zany—and in dazzling colors.

In 1978, 1980, 1984, and 2006, I traveled throughout the highlands of Guatemala researching textiles with butterfly imagery (Ross 2015a). In general, *huipiles* and *cintas* (head coverings), *fajas* (belts), and *tzutes* (utility cloths) often feature realistic butterfly imagery accompanied by flowers and birds. When villagers learned that I was interested in butterfly images, I was often treated to viewing vintage clothing that had been saved from a previous generation or two. On several occasions, the items were offered to me for sale, but in most instances, the textiles were considered family treasures. To quote locals: "Butterflies are both beautiful and sacred, part of our culture."

In 2007, I met a gentleman of pure Mayan descent in Baton Rouge, Louisiana. The individual, named "Jesus," was part of a construction team working at my home. Jesus hailed from a small Mayan village in southern Yucatán, Mexico. Jesus quickly questioned me about the accent tiles in a brick arch within my driveway. The tiles were inscribed "*Na-Pepen*" (a Mayan term that translates into Spanish as "*Casa de las Mariposas*," and into English as "Place of the Butterflies"), and accompanied by a stylized motif of an Aztec butterfly. Delighted that Jesus was literate, I disclosed that I was a butterfly scientist, that I developed my landscape as a sanctuary for butterflies, and that I chose the Mayan term "*Na-Pepen*" based on my past entomological work in southern Mexico and Guatemala. Jesus' eyes glazed over. After a few seconds, he smiled and proudly announced: "Butterflies are sacred to the Maya. In my village there is a damp place where butterflies accumulate. The place is consecrated ground, and we instruct our children not to disturb. We Mayas believe that butterflies transport our spirits after death to the next life." He then stared into my eyes, and with a soft, shaky voice whispered: "You are blessed to be so close to butterflies."

Then, a bombshell. Jesus mentioned that he had been one of the extras that Mel Gibson hired in the filming of his epic 2006 cinematic adventure, *Apocalypto*. "The director," Jesus quipped, "enjoyed learning about Mayan culture and the sacred role of butterflies. Now all Americans can learn about us Mayas."

The continued use of butterfly imagery within the Maya culture bespeaks of contemporary butterfly worship.

Miscellaneous Indigenous Cultures in Mexico

Currently, Mexican government recognizes within its purview sixty-eight living "linguistic groups" with 364 "linguistic varieties" (INALI 2008: 38). From my contacts with linguistic scholars, I can state that no modern culture in Mexico (or Central America) practices "butterfly worship" *sensu stricto*. But because many of these cultures do remain steeped in age-old tradition, individuals often continue to embrace a smattering of ancient beliefs associated with butterflies.

The following are examples of contemporary beliefs about lepidopterans in Mexico—principally from the Trique of Copala and the Mixtec of Magdalena Peñasco (both located in the highlands of Oaxaca) (Hollenbach 1980):

- Butterflies represent the spirits of the dead returning, particularly for "Día de los Muertos." Large colorful butterflies are the souls of adults; the small and white butterflies, the souls of children. Spirits often recruit a living person to accompany them when they return to the afterlife.
- The "God of Thunder" sends the white morpho butterfly (*Morpho polyphemus luna*) (Lamas 2004: 199) to alert people that the seasonal rains are imminent, and that everyone should prepare for floods in ravines.
- When a large butterfly enters a house, someone will die.
- When a butterfly enters a house, the person will go on a journey.
- When little boys drool, providing the pupae of butterflies for the boys to chew will cause the condition to cease.
- The pupae of the mourning cloak butterfly (*Nymphalis antiopa cyanomelas*) (Lamas 2004: 250) are called "sleep insects." When the chrysalises hang from the edge of the roof of a house during their long diapause, the person will have a long sleep.
- The tiger moth (genus *Apantesis*) is referred to as the "butterfly of debt." If one flies into a fire and dies, the owner will owe a lot of money.
- Sphinx (hawk) moths (family Sphingidae) are called "butterfly hummingbirds" because they become hummingbirds.
- Inchworms (larvae of moths within the family Geometridae) are called "measuring worms" because of the insects' loping manner of walking.

Stylized butterfly images are frequently included in contemporary textiles woven on the pre-Columbian style backstrap loom throughout much of Mexico—especially within the Huave culture of coastal Oaxaca. During the winter of 1975–6, and again for shorter periods in 1978, 1980, 1982, and 1983 I traveled throughout southern Mexico in search of Indigenous textiles (Cordry and Cordry 1968) featuring butterflies and moths (Ross 2015a). Although many items were targeted for the frugal souvenir-seeking tourist, some villagers continued to produce vintage designs to entice the more discriminating collector—including those wishing to enjoy a taste of age-old and contemporary butterfly worship (see the previous section on the Mayas).

Zuni Culture of New Mexico, United States

During summer 2006 and autumn 2016, I spent three weeks with the Zuni Tribe (*A:shiwi* or "Flesh"), a Native American culture centered in the Pueblo of Zuni (*Halona Idiwan's* or "Middle Place of the World") in northwestern New Mexico to research butterfly imagery (for reviews, see Ferguson and Hart 1985; Ross 2009, 2015b, 2015c). The Pueblo is the most traditional of the surviving nineteen Rio Grande pueblos in New Mexico. Two residential centers are prominent: Zuni and Black Rock with a combined population of approximately ten thousand. At 6,299 feet (1,920 meters), the Zuni homeland is a semi-arid valley punctuated with monolithic outcrops of sandstone. Anthropologists note that the direct ancestors of the present-day Zuni arrived in the area about 700–800 CE (Ferguson and Hart 1985: 25–32). A Spanish expedition in search of the fabled "Seven Cities of Cibola" was the initial European contact in 1539.

The Zuni follow a form of animism: All things within the landscape are considered spiritual relatives (Erdoes and Ortiz 1984; Sams and Carson 1988; Bruchac 1992; Buffalo Horn Man and Firedancer 1992; Gauding 2006). Animals may be hunted for food but not sport or even scientific collecting. Many animal forms are featured in pottery, silverwork with small inlays of lapidary stones, paintings on canvas, wood, stone, or drum surfaces, beadwork, cotton weavings such as belts, sashes, and mantas (shawls), kachina dolls (small wooden carvings that represent ancestral spirits), and fetishes (small carvings of animals). Foremost, though, are fetishes (Bennett 1993; Rodee and Ostler 1995; McMannis 1998, 2010; Bahti 1999; Ross 2015b, 2015c, 2018).

Fetishes are carved from stone (turquoise is highly prized because of its sacred status), bone, antler, coral, shell, and even juniper wood—all of which are burnished when complete. To outsiders, these carvings seem nothing more than a charm, talisman, amulet, or good-luck piece. Within Zuni framework, however, where every animal is endowed with a unique spiritual nature, fetishes represent a direct connection to the spirit world (Erdoes and Ortiz 1984; Bruch 1992; Buffalo Horn Man and Firedancer 1992; Gauding 2006). The fetish literally encapsulates the animal's spirit, which then becomes more powerful than it was in life. Fetishes, consequently, are not merely emblematic or mythological. They are viable entities that can convey luck, power, protection, and good health in one's everyday experiences.

Zunis recognize two types of fetishes: "ceremonial/true fetishes" and "commercial fetishes." A "ceremonial" fetish is one that has been blessed/consecrated by a spiritual leader. These fetishes must be respected with dignity, and except under unusual circumstances, should not be given or sold to outsiders. Consecrated fetishes may be personal or of a communal nature. If personal, the fetish is worn as a necklace, carried in a separate pouch, or kept in a container that resides in one's residence or vehicle. A communal fetish is often part of sacred tribal ceremony. Ceremonial fetishes are housed in a special "fetish pot" and are "fed" traditional blue corn meal and gifted with crushed turquoise and red coral. By contrast, if a fetish has not been blessed, it is more properly termed a "carving" or "fetish sculpture." It can depart from traditional motifs. It can be given or sold to an outsider as an objet d'art. Many fetishes are accented with inlays of semi-precious stones, but never painted. Today, the term "fetish" is often generalized to include any type of small animal carving.

Butterflies (*Bu:lak'kya*) are highly respected in Zuni culture (Erdoes and Ortiz 1984: 407–8; Bahti 1999: 109–10; Gauding 2006: 129). The butterfly is a "power animal," that is, a spiritual helper, companion, and guide that possesses qualities that humans desire (Gauding 2006: 129). Live butterflies are emblematic for the process of transformation, and the ability to change one's mind and behavior. Because of their bright coloration and graceful movements, butterflies are often endearingly referred to as "dancing flowers" and "air dancers" (Bruchac 1992: 45–7; Buffalo Horn Man and Firedancer 1992: 8; Ross 2015b: 13). Anyone who embraces the butterfly as his/her personal spirit exhibits a heightened sensitivity to the plight of others. "Butterfly People" are noted for their charm, success in business and love, and their sensitivity to the quality of the air (Sams and Carson 1988: 72–4; Gauding 2006: 129; Ross 2015b: 13). Because butterflies are often observed near water and green places, the insects represent the harmony of Earth Mother; an absence of butterflies is a warning that a habitat has been damaged. Finally, when someone sights a butterfly during a contemplative period, the vision confirms the judgment in question. In Zuni tradition, *Bu:lak'kya* are considered the manifestation of Butterfly Maiden (*Bu:lak'kya E:lash'dok'ee*) (Ross 2015b, 2015c), who is represented both in fetishes and kachina dolls. All in all, live butterflies and their representations are benevolent and infuse humans with a sense of wellbeing.

Commercial electric drills, grinders, and polishers are prime tools for creating fetishes. Working the stone generates a lot of dust. The larger particles that accumulate beneath the grinding wheels are periodically returned to the site of excavation; there they are buried with a ceremony. This ritual honors Earth Mother and the spirits of the mountain, and serves as a petition for the spirits to reconstitute the dust into new and more beautiful stone for future carvers. In this way, both spirits and humans are blessed throughout eternity.

On point, too, are kachina dolls. These are small wood-carved representations of kachinas, that is, spirits that control all aspects of the natural world. These doll-like figures are given to children as part of tribal training. Kachinas are worked from the lightweight roots of the cottonwood tree (*Populus deltoides*). When complete, the wood is painted. Zunis recognize a Butterfly Kachina. Each doll is usually five to nine inches in height, and includes recreations of butterfly wings and a butterfly headdress. Water and corn, which are associated with Butterfly Maiden as life forces, are often included.

My host in the Pueblo of Zuni was Edward Lewis (b. 1973), a spiritual leader and gifted artist noted for featuring butterflies in fantasy paintings. Imagine Edward's euphoria when I, a flesh-and-bone butterfly scientist, appeared in the

FIGURE 5.5 Fetish of "Butterfly Maiden" from Pueblo of Zuni, New Mexico. Medium is Zuni stone with inlays of turquoise, red coral, pen shell, and jet dust. Height is 3.5 inches (9 centimeters). Carving by master craftsman Freddie Leekya. Photograph courtesy of Gary Noel Ross.

Pueblo. Edward quickly tagged me with the endearing phrase *Nana Bu:lak'kya* ("Grandpa Butterfly") (Ross 2015b, 2015c). As lagniappe, Edward introduced me to maven carvers, especially those who created butterfly images. With unbridled enthusiasm, Edward commented:

> The idea of painting butterfly maidens came to me when one day in middle school. I was outside eating lunch when I saw two butterflies chasing each other around a flower. My favorite species are the tiger swallowtail [*Papilio rutulus*] and monarch [*Danaus plexippus*], although I sometimes use the queen [*Danaus gilippus*] and mourning cloak [*Nymphalis antiopa*] as well. These are common here on the reservation. I use a butterfly picture book as

a guide for painting details. Because butterflies make me a better person, I hope my images will have that same effect on anyone who gazes on my art. That way Butterfly Maiden will be honored by many outsiders.

Acrylic paints are the medium of choice for Lewis. Each painting is exhaustively detailed; butterflies are anatomically accurate. Each painting, too, is replete with one or more elements of water, maize, marshmallow/rose (*Althaea* sp.), and the hallowed mesa known as *Dowa Yalanne* ("Corn Mountain"). For the final touch, Lewis dusts each work with fine glitter. Says the artist: "Glitter enhances the illusion of mysticism."

During sacred ceremonies, costumed actors portray spiritual kachinas. On the final evening of my initial visit, Edward assembled his dance troop in his back yard to perform the traditional "Butterfly Dance." The performance was solemn and spanned several hours. At the end, we partook of a traditional feast that included yellow cornmeal sweet bread and squash seed stew. The next morning, I joined Edward at his residence. When the first rays of the sun broke the horizon, Edward addressed Butterfly Maiden in his native tongue. Then Edward assured me that I, *Nana Bu:lak'kya*, would be protected for my return to Louisiana. And so I was.

During my second visit (2014), I was again graced by Butterfly Maiden. Shortly after arriving in the Pueblo, I slipped and hit my head. Because I began to lose motor skills, I was quickly medevacked by helicopter to the University of New Mexico Hospital in Albuquerque. There I underwent cranial surgery to remove what doctors diagnosed as a "massive subdural hematoma." After twelve days in the ICU, I was released by the surgeon who stated that my recovery was nothing short of "miraculous." I returned to the reservation and spent the following two weeks as a guest of the Edward Lewis family so that I could continue my research (Ross 2015a, 2015c). Edward was of the opinion that my accident was the result of a malevolent spell inflicted by a diabolical person. Nevertheless, he stated: "*Bu:lak'kya E:lash'dok'ee* is now in control and will protect her *Nana Bu:lak'kya*." Each night before bed, Edward performed a short sacred ritual to entreat *Bu:lak'kya E:lash'dok'ee* to protect everyone within his home from malevolent spirits. When it was time for me to begin the 1,500 mile (2,416 kilometer) drive to my residence in Baton Rouge, Edward was resolute: "*Bu:lak'kya E:lash'dok'ee* will continue to protect you." And once again, she did (Ross 2015c: 78).

Between January 2008 and December 2009, I sponsored a major exhibit in the "Hall of Culture and Science" at the McGuire Center for Lepidoptera and Biodiversity, Florida Museum of Natural History, University of Florida (Gainesville) (Ross 2009). The display consisted of eighty-five pieces and was titled "Butterflies and Moths in Contemporary Zuni Art."

EPILOGUE

Within virtually all industrialized societies of this twenty-first century, live butterflies and their images are alluring and pervasive (Ross 1996; Pyle 2001; Boppré and Vane-Wright 2014). Summarily: 1) butterfly/pollinator gardens are popular in residential/commercial landscapes and public areas. Installations usually include milkweeds, the unique reproductive plants of the monarch butterfly; 2) teachers often employ commercial kits of live butterflies to demonstrate metamorphosis; 3) butterfly conservatories/houses and theme parks exhibiting collections of live butterflies within lush garden settings are increasingly common; 4) live butterflies are frequently released at social ceremonies such as weddings, funerals, birthdays, and at the dedications of buildings, parks, bridges, and ships; 5) glamorous images of butterflies (real and fantastic) commonly decorate miscellaneous fabrics, clothing, footwear, jewelry, body art, assorted knickknacks and home décor, and in addition, appear in movies, television programs, and media advertisements; 6) the butterfly image is the logo for organizations that inspire hope and change such as Alcoholics Anonymous, Compassionate Friends, and Hospice; 7) the butterfly image often illustrates banners, flags, and printed material by Christian sects (especially Catholic, Episcopal, Lutheran) on Easter Sunday to symbolize the resurrection of Jesus Christ; 8) the monarch butterfly is acknowledged as the "de facto national butterfly" for the United States; 9) and finally, because of the increasing beleaguered status of the monarch butterfly, the species serves as a "powerful metaphor ... and stands for every fine and delicate thing that philistine force and greed would gladly eradicate from this bruised and beautiful world" (Pyle 2001: 25).

Entrepreneurs have been quick to take advantage of this enhanced market for live butterflies; butterfly husbandry involving specialized ranches and farms has become lucrative (Ross 1996; Pyle 2001: 21; Boppré and Vane-Wright 2014). Pupae and freshly enclosed adults are shipped around the world. Butterfly husbandry has sparked a "spiritual awakening" within the public—at least within the United States and Great Britain.

But there is a downside. Whereas breeding butterflies for conservatories is generally accepted by professional organizations (North American Butterfly Association and The Lepidopterists' Society, for example), the release of laboratory-bred specimens into the natural environment is not (Glassberg et al. 1998). (Remember the adage, "Don't mess with Mother Nature.")

The butterfly's unrivaled loveliness, sensitivity to the environment, seemingly unworldly life cycle, and widespread presence in mythology and religion all echo across time and space. Now, in the twenty-first century, butterflies continue to entrance, charm, comfort, and seduce us. Put another way, our sense of awe from butterflies heals our souls. But to champion this as a renaissance of animism may, for some, be a bit presumptive. Names aside, though, such distinction would not be too different from the veneration we lavish on Christian, Jewish, and Muslim saints and prophets. History tells us that humanity is best served when our psyche is free to imagine, to ramble through the unknown without having to demystify everything. This psyche also cries out for therapeutic skills and representations—pious or otherwise—that have the power to grant freedom from pain and sorrow.

In 1991, Miriam Rothschild penned the following in response to the popularity of books on butterflies in her native land: "Perhaps the explanation ... lies not in the highly commendable mushrooming of interest in the fauna of the EEC, but in the remnant of butterfly worship that lingers within us all" (1991: 52). I agree. Furthermore, I posit that the ubiquitous mantra of *Homo sapiens* to transcend from the physical world stems from a unique neurological pathway that is hardwired in our DNA—the molecular blueprint for all that we are. What initially must have been nothing more than esoteric peculiarities that were catalyzed by random genetic variations were, little by little and generation upon generation, winnowed by the forces of natural selection into cumulative behavior that improved the functioning and survival of our early ancestors. Eventually, this behavior became a defining and ennobling trait of our species.

To reflect: As far back as the nineteenth century, Charles Darwin argued that aesthetics can result from natural selection (see discussions in Dutton 2008: 85–6, 164–5, 204, 206–7; Hammond et al. 2017: 34; Prum 2017: 17–55), and therefore can convey an evolutionary advantage. More recently, Denis Dutton has theorized that our love of beauty is inborn, and that many so-called artistic expressions are systemic across cultures (2008). Richard Prum has pushed the envelope further by arguing for an "aesthetic view of evolutionary biology" where he defines beauty as "the result of a coevolutionary dance between desire and display" (2017: 320–49). Such encoding of our genome expresses itself in several ways: our universal fixation on art, color, dance, bucolic landscapes, music, poetry, trees and flowers—and of course, butterflies. All make us feel good, lead happier lives, and so, increase our longevity.

In conclusion, the modern butterfly paradigm—extenuated as it may be—is an inexorable revival of ancient *butterfly worship*. The concept seems genuinely plausible, axiomatic, and it does not require a literal meaning tied to a formal practice. To be candid, I remain passionate about butterflies. They are sobering, gratifying to my psyche. And because I am not only a scientist but a man of the Faith as well, butterflies offer me a preview of a glorious hereafter—a glimpse of Nirvana, the Divine. But whether one is a theist, atheist, animist, Darwinian humanist, or anything in between, the butterfly inspires, affirms, and unifies our humanity. As such, the butterfly has been, and continues to be, a unique blessing from the Cosmos.

CHAPTER SIX

Insects as Symbols

Insects as Tricksters in Early Blues Music

ROBERT K.D. PETERSON

Modernity—with its reliance on precise descriptions of phenomenological reality—has not lessened the importance of insects as symbols, but their symbolic nature has evolved to reflect the modern age. Instead of insects portending supernatural evil or the capriciousness of gods and monsters—as in previous ages—they have in many instances become symbols of runaway science and technology. Conversely, instead of insects portending supernatural benefit or good luck, they have in many cases become symbols of healthy ecosystems and natural balance.

Giant insects portrayed in film and literature in the 1950s and 1960s were symbols of fears of communism and out-of-control technologies. Many of the insects grew to their giant size because of mutational exposure to radiation, which was culturally a source of fascination and fear. In arguably the best giant-insect film of the period, *Them!*, ants become huge and terrorize the residents of Los Angeles. The ants are symbols of "ruthless, destructive enemies of the state" that must be vanquished. Science is responsible for their monstrous size, but ironically science must be used to solve the crisis (see Chapter 3 in this volume).

Insects as symbols of the balance of nature and natural harmony are a relatively recent phenomenon. Butterflies have long been used as symbols of natural purity, but only in the last decade have bees emerged as "poster children" for environmental concerns and activism. In the 1970s, society was encouraged to "save the whales"; now we are encouraged to "save the bees."

Bees are a cultural icon and touchstone for all things environmental because of recent concerns about shrinking populations of some bee species (see Chapter 4). The recent trend of using bees as symbols for natural balance shows no signs of abating as demonstrations, environmental justice concerns, sustainability, and marketing have combined to elevate the bee as a major cultural symbol.

With the rise of computers in the last two decades of the twentieth century to the present, the terms "computer bug," "software bug," "debug," or just "bug" are arguably the most popular or familiar contemporary terms with insects as symbols for something else. However, the origin of "bug" to represent a computer glitch is misunderstood by most. The use of "bug" to represent a defect predates the electronic computer by decades. Indeed, it has been used in engineering since at least the 1870s. In 1878, Thomas Edison wrote:

> The first step is an intuition, and comes with a burst, then difficulties arise—this thing gives out and [it is] then that "Bugs"—as such little faults and difficulties are called—show themselves and months of intense watching, study and labor are requisite before commercial success or failure is certainly reached.
>
> (Hughes 1989: 75)

The term "bug" is derived from the Middle English "bugge," which is the source of "bugbear" and "bugaboo." Those terms mean "monster." So, "bug" as a term to describe a glitch in a system originally was meant to mean "bugaboo," a monster. The modern use of "bug" to describe computer and software glitches definitely has a strong insect connection. In 1947, Grace Hopper, a computer engineer, noted that computer operators traced an error in a computer to a moth trapped in a relay. The moth was removed and taped to a log book. The operators who found it were familiar with "bug" as an engineering term and affixed the moth to the log book with the note, "First actual case of bug being found." The log book is now at the Smithsonian National Museum of American History.

***

Insects are invoked in numerous contexts in popular music in the modern age, but in many cases they are not symbols for anything. What is the symbolic significance of the beetle in the band name The Beatles? There is no insect symbolism in what is arguably the most important popular music group of the modern age. The Beatles' name is a take-off on the Crickets, Buddy Holly's band. The first name in 1960 was the Beetles, but John Lennon cleverly transformed the band name from one referring to the coleopteran insect to one referring to the then popular beat poetry movement, hence The Beatles.

Before finally settling on The Beatles in the second half of 1960, the band went through various related names, such as The Silver Beatles and The Beatals. So, the most famous and significant popular music group has an insect name not only of no symbolic significance, but it was also changed so as to not refer to actual beetles.

Other references to insects in popular music in the modern age are steeped in symbolism. In the rock-and-roll era, the song (also the album title) "Maggot Brain" by Funkadelic, released in 1971, uses "maggots in the mind of the universe" to represent human-caused environmental decay. The song "Cuckoo Cocoon" from the concept album *The Lamb Lies Down on Broadway* by Genesis, released in 1974, symbolizes profound changes by the lead character, Rael (Coelho 2000). Although Clutch's song "The Dragonfly," released in 1998, is a full-on paean to nature, especially insects and the rite of spring, Ziggy Marley's song "Dragonfly," released in 2003, uses the dragonfly to represent nature's judgment of humans: "Hey Miss Dragonfly I see you look at me with your beautiful eyes | You must be wondering what type of creature am I."

Jethro Tull's song "Moths," released on the album *Heavy Horses* in 1978, speaks to moths as young lovers and spring's renewal of life: "And the first moths of summer | Suicidal came, suicidal came."

Conversely, the song "Moth" by Audioslave, released in 2006, uses the moth in an Icarian tale of defeat: "I was a king, I was a moth with painted wings made of cloth | When did the flame burn so high and get so hot?"

The song "Bug" by Wussy, released in 2014, uses the insect to symbolize personal weakness and demons: "You're the drink, you're the drug | You're the bug that's alive inside of me."

***

Whether symbols or not, examples of insects in lyrics, songs, albums, and musicians' names in the modern age are too numerous to list. One genre with compelling uses of insects as symbols is blues music in the United States in its early period. These uses are culturally important because they open a window on US history to reveal a host of uncomfortable social topics.

The blues is arguably the oldest and deepest of all music unique to the United States, but it was not created out of whole cloth. Like the melting pot that is America, blues was derived from several diverse sources. These include field hollers, gospel, European folk music, rag, and African scales and rhythms. These diverse influences converged around the turn of the twentieth century to form blues music.

W.C. Handy, bandleader and composer, was waiting for a train at the Tutwiler, Mississippi depot about 1903 when he saw a man playing guitar using a knife as a slide in a distinctive style (Mississippi Blues Commission 2003). The man was singing a song about "going where the Southern cross' the Dog."

Handy wrote:

> A lean loose-jointed Negro had commenced plunking a guitar beside me while I slept ... As he played, he pressed a knife on the strings of the guitar in a manner popularized by Hawaiian guitarists who used steel bars .... The singer repeated the line three times, accompanying himself on the guitar with the weirdest music I had ever heard.

Using this experience as inspiration, Handy later wrote out the musical notation of the twelve-bar song and published it in 1912 as "Yellow Dog Blues." Through publishing and performing this song and many others, Handy, if needed was crucial to popularizing the blues as a musical genre (1941: 74).

As early as 1909, "blues" as a term for the musical style was being used and understood by the public. As a result of meticulous historical research, Abbott and Seroff (2017: 4) argued that 1909 was the year the blues became a distinct and recognizable musical genre. Around this time, the blues migrated from being played by individuals and small combos on and near southern plantations to being featured music on the black vaudeville stage throughout the country (Abbott and Seroff 2017: 4). As vaudeville was fading, the blues was being recorded and became even more popular as phonographs became omnipresent household possessions.

The music of the blues was not the only thing not created out of whole cloth. The lyrics of the blues borrowed from many sources. A very important African source for lyrics was the animal trickster, a figure in African and African American narrative tradition that "flouts the norms of society, using cunning, humor and deceit to obtain personal gain" (Smith 2005). Perhaps the most recognizable animal trickster is Br'er (Brother) Rabbit, a hare from African storytelling traditions, although in West Africa the stories that have come to be associated with Br'er Rabbit happen to the spider Anansi (Opala 1987: 36).

In American folklore, Br'er Rabbit uses his cunning to escape from the clutches of Br'er Fox. However, like many animal tricksters, Br'er Rabbit can be a protagonist or antagonist or both, depending on the story (Arnold 1996: 290). The stories can be complex, with the trickster protagonist also exhibiting amoral and anti-social behavior. These stories are meant to convey that sometimes people must exhibit odious behaviors to survive, but those behaviors are not necessarily to be applauded. Levine (1977: 560) argued that in the American stories the animal trickster represented the slave who must use cunning to survive and exact revenge.

The animal trickster is represented well in early blues lyrics (Smith 2005). In particular, there are numerous songs and lyrics with insect tricksters. In the following three sections, insects as tricksters will be discussed in some detail.

CHARLEY PATTON AND THE BOLL WEEVIL

Charley—sometimes spelled "Charlie"—Patton's influence on twentieth-century music is impossible to overstate. He helped found and solidify the Delta blues tradition at the beginning of the twentieth century. The Delta blues is unarguably the most important blues form and jazz and rock grew directly out of it.[1]

Born April 1891 near Edwards, Mississippi, Patton moved with his family to the giant Dockery Plantation in the Delta region of the state about 1900. There, he and his family worked cotton fields like so many other African American families during that time. "Even though the work was backbreaking

FIGURE 6.1 The only known photograph of Charley Patton, 1929. WikiCommons Media, https://commons.wikimedia.org/wiki/File:Charley_Patton_(1929_photo_portrait).jpg.

and unforgiving, there were vibrant communities in the Mississippi Delta. Juke joints, plantation dances, fish fries, picnics, and all-night frolics were social gathering places and events for plantation workers who needed to be entertained" (Peterson 2007). Patton was more than happy to entertain.

As a boy, he learned to play guitar from Henry Sloan, an older plantation tenant, about whom virtually nothing is known. Patton created his music by combining field-holler vocal stylings, percussive acoustic guitar (reminiscent of West African polyrhythmic drumming), and high-note accents on the guitar using a bottleneck or knife blade to slide across the fretboard (Peterson 2007). Into this mix, he added lyrics that directly reflected the southern African American experience at the time.

> Listening to it, you sense the sticky mud and searing summer heat of the bottomland, the tenant shacks, and eerie spectres at desolate crossroads. You feel the despair, joy, hate, fear, wanderlust, and heartbreak. To put it simply, the Delta blues is the truth. That's why it resonates so deeply across time and culture.
>
> (Peterson 2007)

FIGURE 6.2 "King Cotton." Panoramic gelatin silver photograph by J.C. Coovert, 2007. US Library of Congress Prints and Photographs Division, Washington, DC [pan 6a36702].

Patton was a popular performer who entertained audiences throughout the Delta. His loud, gruff voice and guitar showmanship left audiences spellbound and begging for more. He was arguably the first to play his guitar behind his head and between his knees, imitated to great effect by rock legends Chuck Berry and Jimi Hendrix.

Peterson (2007: 143) wrote:

> Although he lived much of his life on the Dockery Plantation, he set up households with different women all over the Delta. His life was stereotypical of a traveling bluesman. He drank heavily, chain smoked, and was an inveterate womanizer. Small-town jails were second homes to him, and he always had to keep an eye out for jealous men.

In the late 1920s and before the Great Depression, record companies were in great competition to record all things blues and Patton was invited to record some of his songs. On June 14, 1929, Patton cut fourteen songs at Gennett Recording Studios in the unlikely location of Richmond, Indiana. The songs were soon released as 78-rpm records on the Paramount label. Among the songs was "Mississippi Boweavil Blues" (Peterson 2007).

Patton's "Mississippi Boweavil Blues" is an archetypal blues song. Rather than being in the familiar twelve-bar format, it is primal blues—one chord accompaniment, three basic notes in the vocal melody, and a high-note bottleneck accent after the vocal phrase (Peterson 2007). The lyrics are remarkable because Patton is singing about an actual insect—rather than the insect serving as a sexual or other metaphor—and using the insect as a trickster:

Sees a little bo weevil keeps movin' in the air, Lordie!
You can plant your cotton and you won't get half a bale, Lordie
Bo weevil, bo weevil, where's your native home, Lordie
"A-Louisiana raised in Texas, least is where I was bred and born," Lordie
Well I saw the bo weevil, Lord, a-circle, Lord, in the air, Lordie
The next time I seed him, Lord, he had his family there, Lordie
Bo weevil left Texas, Lord, he bid me "fare ye well," Lordie
(Spoken: Where you goin' now?)
I'm goin' down the Mississippi, gonna give Louisiana hell, Lordie
(Spoken: How is that, boy?)
Suck all the blossoms and he leave you half a square, Lordie
The next time I seed you, you know you had your family there, Lordie
Bo weevil meet his wife, "We can sit down on the hill," Lordie
Bo weevil told his wife, "Let's trade this forty in," Lordie

Bo weevil told his wife, says, "I believe I may go North," Lordie
(Spoken: Hold on, I'm gonna tell all about that)
"Let's leave Louisiana, we can go to Arkansas," Lordie
Well, I saw the bo weevil, Lord-a-circle, Lord, in the air, Lordie
Next time I seed him, Lord, he had his family there, Lordie
Bo weevil told the farmer that "I 'tain't go ticket fare," Lordie
Sucks all the blossom and leave you half a square, Lordie
Bo weevil, bo weevil, where your native home? Lordie
"Most anywhere they raise cotton and corn," Lordie
Bo weevil, bo weevil, "Oughta treat me fair," Lordie
The next time I did you had your family there, Lordie

Patton may have been performing this tune as early as 1910, making it among the first group of blues songs. By that time, the boll weevil would have been known to nearly everyone in the deep South. Since entering the United States near Brownsville, Texas in 1892, it moved rapidly over the next thirty years to infest six hundred thousand square miles. In 1907, the first boll weevil was found in Mississippi, and by 1915 populations covered the entire state. The social and economic upheaval wrought by the boll weevil is legendary. Cotton crops failed, land prices sank, farmers went bankrupt, and farm workers fled to industrial cities in the north (Peterson 2007).

Patton undoubtedly saw all of this and his lyrics reveal the devastation and inexorable invasion of the pest as it moved from Texas through Louisiana and Mississippi.

> The lyrics reflect what he saw directly: tremendous upheaval and uncertainty for the plantation owner, and, especially, for the African-American tenants. Yet, at the same time, there is envy for the boll weevil in the lyrics. The boll weevil could go anywhere it wanted; the poor tenant farm worker could not. The boll weevil had power over the plantation owner; the farm worker did not.
>
> (Peterson 2007)

Patton's song evolved throughout the next several decades. Ma Rainey and Bessie Smith sang a different song about the boll weevil. It doesn't seem to be related to Patton's tune in music or lyrics. The boll weevil here is clearly a no-good man: "I don't want no man; to put sugar in my tea; that bug is so evil; I'm afraid he might poison me."

The other boll weevil songs can be traced directly to Patton's version. Other artists have added greatly to the lyrics and it has become much more melodic over the years. Sometime during the late 1930s, the version was transformed into a traditional twelve-bar blues piece with jaunty, humorous lyrics that even included a reference to Paris green, the arsenical insecticide. Not only are there blues versions by Lead Belly, Pink Anderson, Little Walter, and Jaybird Coleman, but there also are folk versions by Woody Guthrie, Burl Ives, and Pete Seeger, a jazz version by Eubie Blake, a bluegrass/country version by Tex Ritter, R&B versions by Brook Benton and Fats Domino, and rock versions by Eddie Cochran and Jimmy Page. Many of the versions were recorded during the folk revival movement of the late 1950s and early 1960s (Peterson 2007). The titles of the songs varied; examples include "Bo-weavil Blues," "New Bo Weavil Blues," "Boll Weevil Rag," "The Boll Weevil Song," "Boll Weevil," "Let Me Be Your Bo Weavil," and "Just Lookin' for a Home." A Second World War adaptation by Pete Seeger was called "Looking for a Home."

But it all started with Charley Patton. His song seems to be the original and has the deepest links to the historical boll weevil invasion and the devastation it caused (Peterson 2007). He recorded several more records for the Paramount and Vocalion labels in 1930 and 1934, and, among the so-called "race" records, his 78s were always strong sellers. But, he was not long for the world. At age forty-two on April 28, 1934, with his common-law wife, Bertha Lee, tending to him, Patton died of a mitral valve disorder in his heart. He is buried in Holly Ridge, Mississippi—near an old cotton gin.

BLIND LEMON JEFFERSON AND THE MOSQUITO

On September 24, 1929, Blind Lemon Jefferson recorded "Mosquito Moan," a three-minute song for Paramount Records at Gennett Recording Studios in Richmond, Indiana. Unfortunately, the song was recorded at his final recording session. He died shortly after at the age of thirty-six.

Lemon Henry Jefferson was born on a farm in the now abandoned town of Couchman, near Wortham, Texas. According to census records, he was born on September 24, 1893, but he indicated he was born on October 26, 1894 on his 1917 draft registration (Peterson 2013). His parents were sharecroppers and he was but one of eight siblings. In his First World War draft registration, he revealed that he was blind since birth, but the extent of his blindness is unclear as the best-known photo of him shows him wearing eyeglasses.

In his early teens, Jefferson began playing guitar and performing as a street musician in several East Texas towns. During the early 1910s, he often traveled north to Dallas to play with future blues giants Lead Belly and T-Bone Walker—the latter was taught guitar by Jefferson in exchange for his services as the blind musician's guide.

A recording scout discovered Jefferson in 1925 and soon thereafter he traveled to Chicago to make records with Paramount Records. He recorded more than a hundred sides from 1926 to 1929 and became a nationally known

FIGURE 6.3 Photograph of Blind Lemon Jefferson. Photograph courtesy of GAB Archive/Redferns/Getty Images.

blues artist while he was alive (Peterson 2013). He was among the first group of blues artists to record solo voice and guitar accompaniment. Among his hits were "Matchbox Blues," "See That My Grave Is Kept Clean," and "Black Snake Moan." His fingerpicking guitar style was intricate for that time and his vocals were often high-pitched and loud (Peterson 2013). He was fond of women and whiskey, but refused to play the blues on Sundays—evidence of the constant struggle between blues and religion among many musicians at that time.

The song "Mosquito Moan" was released in January, 1930 on the flip side of the 78-rpm record with his song "Southern Woman Blues."

I'm sittin' in my kitchen, mosquitoes all around my screen
I'm sittin' in my kitchen, mosquitoes all around my screen
I'm about all ready to get a mosquito bomb, I'll be seldom seen

I must sleep under a tin tub, try my best to brad their bills
I believe I'll sleep under a tin tub, try my best to brad their bills
Mosquitoes so bad in this man's town, keep me away from my whiskey still

I love my whiskey better than some people likes to eat
I say I love my whiskey better than some people likes to eat
Mosquitoes bother me so I can hardly stay on my feet

I bought a spray last night and I sprayed all over my house
I bought a spray last night 'n I sprayed all over my house
Mosquitoes all around my door won't leave nobody come out

Mosquitoes all around me, mosquitoes everywhere I go
Mosquitoes all around me, mosquitoes everywhere I go
No matter where I go, they sticks their bill in me

I would say gabbernipper, these gabbernippers bites too hard
I would say gabbernipper, some gabbernippers bites too hard
I stepped back in my kitchen and they springin' up in my back yard

The first verse sets the stage. Jefferson is trapped in his house and he's contemplating drastic action to rid himself of the vexing trickster mosquitoes. The second verse uses the term "brad their bills" which has been misinterpreted by some as "grab their bills." The term "bill" refers to the female mosquito's proboscis, or sucking mouthparts. The term "brad their bills" means that Jefferson hopes the tin tub he's sleeping under will bend their bills so they will no longer be able to suck his blood. Here, "brad" is in verb form referring to the small nail with an irregular head. The end of the second verse and the entire third verse refer to Jefferson's love of hard liquor at the height of Prohibition. We learn why he is so bothered by mosquitoes: they are preventing him from getting to his whiskey still. The fourth verse has him spraying for mosquitoes in his house, but the last line tells us it is futile because many more are waiting outside his door. The fifth verse reinforces the fact that no matter how hard he tries, the mosquitoes—as tricksters—will continue to vex him. The sixth and final verse uses the term "gabbernipper," which is not commonly used today. "Gabbernipper" or "gallinipper" is a southern American term for a large mosquito or crane fly.

Peterson (2013: 112) wrote:

> We will never know why Jefferson chose to write a song about mosquitoes. Other blues artists at the time were singing about bees, bed bugs, and boll weevils, but not mosquitoes. On the surface, the lyrics are humorous and only seem to be about Jefferson's battle with mosquitoes. But, like many of the blues songs of the period, there is something else going on here. The mosquito in the song is a trickster, an animal figure often used in African American story and song of the time to allude to cunning, humor, and deceit

to obtain personal gain. In addition, the mosquitoes are metaphors for human tormentors, constantly dogging Jefferson as he tries to live his life in early twentieth century America. This is most apparent in the fifth verse, "Mosquitoes all around me, mosquitoes everywhere I go; No matter where I go, they sticks their bill in me."

Unlike many of his other songs, "Mosquito Moan" has rarely been covered by other artists. Blues artist Paul Geremia covered the song on his 2004 album *Love, Murder & Mosquitos*. Unfortunately, Jefferson did not live to see the release of his song. He died in Chicago on December 19, 1929, most likely of a heart attack after becoming disoriented during a snowstorm. Paramount Records paid for his body to be returned to Texas by train and he was buried at Wortham Negro Cemetery. The name of the cemetery was changed to Blind Lemon Memorial Cemetery in 2007 (Peterson 2013).

FURRY LEWIS AND THE BED BUG

The story of the song "Mean Old Bedbug Blues" is tied inextricably with the blues musician Furry Lewis. Walter E. Lewis was born in Greenwood, Mississippi on March 6, 1893. At the young age of seven, he moved with his family to Memphis, Tennessee, where he would live for the rest of his life. As a child, his friends called him "Furry" and the nickname stuck. Shortly after arriving in Memphis, Lewis learned the guitar from a middle-aged man he called "Blind Joe." At the young age of fifteen, Lewis was playing for street corners, bars, and private parties.

A stint with W.C. Handy's Orchestra led to several years of him playing on the road. In 1917, Lewis lost a leg in a railroad accident and returned to Memphis to play solo and with various small blues combos. He quit touring in 1922, returned to Memphis, and took a steady job as a street sweeper (Peterson 2014).

However, Lewis stayed active in music. In April and October 1927, he recorded his first eleven records, known as "sides," in Chicago for the Vocalian label. Then, he recorded for the Victor label in Memphis before recording for Vocalian again in 1929. Even though some of his records, like "Kassie Jones," are now considered blues and folk classics, none of his records sold well at the time, unlike those of Charley Patton and Blind Lemon Jefferson.

Among the songs he cut in 1927 was "Mean Old Bedbug Blues," a two-minute, forty-three-second traditional twelve-bar blues song with humorous lyrics. On some of the record labels for the song, the songwriter is credited as

Leo Wood, although it is unclear if he wrote the music, lyrics, or both (Peterson 2014). Regardless, Lewis recorded the most distinctive version of the song with only his guitar to accompany his signing. He plays the guitar on the song in a light, lilting manner and delivers the lyrics in a no-nonsense style with his strong singing voice, which was his forte.

Mean old bedbug sho' is evil, he don't mean me no good
Mean old bedbug sho' is evil, he don't mean me no good
He thinks, "I'm a woodpecker," and he takes me for a chunk o' wood

When I lay down at night I wonder how can a poor man sleep
When I lay down at night I wonder how can a poor man sleep
With one holdin' your hand while the other one eatin' your feet

Bedbug's big as a jackass, he will bite you, stand and grin
Bedbug's big as a jackass, will bite you and stand and grin
Drink a bowl of bedbug poison, come back and bite you again

Saw one moanin' in a corner, Lord I tried so hard to see
Saw one moanin' in a corner, Lord I tried so hard to see
It was a mother bedbug, Lord, prayin' for some more to eat

I had to sit up all night long, my feet can't touch the floor
I had to sit up all night long, my feet can't touch the floor
Cause the mean old bedbug told me I can't live there no more

Other musicians played and recorded "Mean Old Bedbug Blues," also titled "Mean Old Bed Bug Blues," "Big Bed Bug," "Bedbug Boogie," and several other variations on recordings. Other artists of the era to record it were Tommy Settlers and His Blues Moaner, Lonnie Johnson, Lizzie Miles, Betty Gray, Kitty Walters, Bobby Leecan and Robert Cooksey, and Fats Waller with Jack Bland and His Rhythmmakers. Lonnie Johnson recorded not only "Mean Old Bed Bug Blues," but also a sequel, "Bedbug Blues, Part 2." Later recordings were from Lightnin' Slim, Earnest Tubb, Ramblin' Jack Elliott, Dave Van Ronk, and Suzy and Eric Thompson.

The best-known recording of the song, however, was by Bessie Smith on September 27, 1927, in which she was accompanied only by a guitar and piano. Many versions of the song, especially those recorded by blues women, can be traced directly to Smith's three-minute, eleven-second version.

Smith was a major star in her time and popular with both black and white audiences. Her huge voice and larger-than-life personality earned her the title,

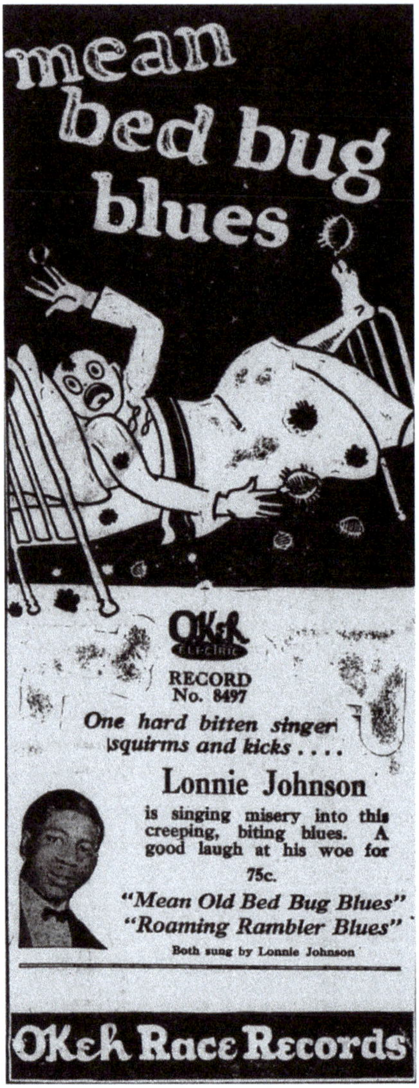

FIGURE 6.4 1927 advertisement for Lonnie Johnson's recording of "Mean Old Bed Bug Blues" in *The New York Tribune*. Courtesy of Library of Congress.

"Empress of the Blues." She was born on April 15, 1894 in Chattanooga, Tennessee and lived an amazing—though short—life that has been well chronicled.

Smith's parents died when she was very young and she and her siblings raised themselves in desperately poor circumstances. She sang and danced for spare change on the streets of her hometown before leaving in 1912 to join a vaudeville troupe. Her strong voice and the surprising popularity of recorded blues music, also known as race records, in the early 1920s led to her recording

FIGURE 6.5 Photograph of Bessie Smith. Photograph courtesy of Michael Ochs Archives/Getty Images.

contract. She recorded 160 sides for Columbia Records from 1923 to 1933. Her records and live performances made her the highest paid African American entertainer of her day (Peterson 2014).

Gals, bed bugs sure is evil, they don't mean me no good
Yeah, bed bug sure is evil, they don't mean me no good
Thinks he's a woodpecker and I'm a chunk of wood

When I lay down at night, I wonder how can a poor gal sleep
When I lay down at night, I wonder how can a poor gal sleep
When some is holding my hand, others eating my feet

Bed bug as big as a jackass, will bite you and stand and grin
Bed bug as big as a jackass, will bite you and stand and grin
Will drink all the bed bug poison turn around and bite you again

Something moan in the corner, I went over and see
Something moan in the corner, I went over and see
It was the bed bug a prayin', Lord, gimme some more to eat

Got myself a wishbone, bedbugs done got my goat
Got myself a wishbone, wish they'd cut their own doggone throats

Although her lyrics are very similar to Lewis's, Smith's version is more clearly a double-entendre affair compared to Lewis's bed bug squarely in a trickster role. Smith's bed bug is a metaphor for troublesome men. The sexual imagery is palpable and far from subtle (Peterson 2014).

The lyricist and the singers and musicians of the era who performed the song probably had firsthand experiences with bed bugs and therefore had the motivation to sing passionately about them. Like now, they were difficult to get rid of once they found their way into a dwelling. Until the use of DDT in the mid-1940s, control methods such as oil sprays, pyrethrum sprays, and sulfur dioxide fumigations were ineffective (Peterson 2014).

Furry Lewis continued working for the City of Memphis until he retired in 1966. He never gave up music, though, and was part of the blues revival movement of the 1960s. He was recorded several times in the 1960s and early 1970s and he was in an eclectic mix of activities before he died, including opening twice in concert for The Rolling Stones, playing on the *Tonight Show*, being featured in *Playboy* magazine, and appearing in the Burt Reynold's movie, *W.W. and the Dixie Dancekings* (Peterson 2014). In 1976, Joni Mitchell wrote and recorded a song about him, "Furry Sings the Blues," that was not flattering to Lewis and he was not happy with Mitchell for recording it. He gradually lost his eyesight, contracted pneumonia, and died of heart failure in 1981.

The Great Depression greatly curtailed Bessie Smith's career, but she continued to work until she died from injuries caused by an automobile accident near Clarksdale, Mississippi on September 26, 1937. She was forty-three. She is recognized today as a member of an elite group of the truly great singers of the twentieth century, and her influence can be heard in many vocalists that have come since.

With the resurgence of bed bugs and renewed attention to them in recent years, the song "Mean Old Bedbug Blues" causes us to pause and reflect on how these insects were perceived nearly ninety years ago. Life in the United States is very different than it was in 1927, but the bed bugs are still with us.

INSECTS AS SEXUAL METAPHOR

Although Charley Patton's "Mississippi Boweavil Blues," Blind Lemon Jefferson's "Mosquito Moan," and Furry Lewis's "Mean Old Bedbug Blues" were about actual insects, they also were clearly metaphors for troublesome people. Nearly all other blues songs that feature insects are overt metaphors for sex and sexual prowess.

Bo Carter's "I'm an Old Bumble Bee" is a typical example:

I am an old bumblebee
A stinger just as long as my arm
I stings every good-looking woman now
Everywhere I goes along

The song titles are too numerous to list here and some songs are so old and have been covered by so many artists that it is nearly impossible to attribute songwriters to them, but the following examples will suffice. "Cockroach Blues," "Chinch Bug Blues," and "Creeper's Blues" continue the bed bug blues theme, except the insects are different (the creeper refers to the cockroach). Blind Lemon Jefferson recorded "Chinch Bug Blues" and Furry Lewis recorded "Creeper's Blues."

The most interesting examples of insects as tricksters and sexual metaphors are the ants, wasps, and bees—collectively known as the Hymenoptera. "The obsession in blues songs with the stinger as a phallic symbol and making honey as a sign of male virility is ironic and comical given that stingers and honey-making are exclusive to female hymenopterans" (Peterson 2007). Clarke (2014) discussed the history of bees on record labels.

Bo Carter's "Ants in My Pants" has nothing to do with ants in his pants, and the lyrics are sure to disappoint a myrmecologist. Bee and wasp examples include the aforementioned "I'm an Old Bumble Bee," "Yellow Jacket," "Bumble Bee," "Bumble Bee Blues," "New Bumble Bee," "Honey Bee," "I'm a King Bee," "Queen Bee," "Yellow Bee," "Little Queen Bee," "Back Biting Bee Blues," "Little Bee," and "Stinging Bee Blues." Memphis Minnie's partial lyrics to "Bumble Bee" are typical of the bee blues songs: "Bumble bee, bumble bee, please come home to me | He got the best old stinger any bumble bee that I ever seen."

Muddy Waters's lyrics for "I'm a King Bee" are typical of the often-covered song:

Well I'm a king bee
Buzzing around your hive
Well I'm a king bee, baby
Buzzing around your hive
Yeah I can make honey baby
Let me come inside

Well I'm a king bee
Want you to be my queen
Well I'm a king bee, baby

Want you to be my queen
Together we can make honey
The world has never seen

Well, buzz a while
Sting you bad

Well I'm a king bee
Can buzz all night long
Well I'm a king bee, baby
Can buzz all night long
Yeah I can buzz better baby
When your man is gone

Not all arthropods in blues lyrics are insects. Spiders are well represented by "Spider Blues," "Spider Bite Blues," "Spider Man Blues," "Spider Web Blues," "Spider Woman Blues," "Black Spider Blues," and "Mean Red Spider." Muddy Waters's partial lyrics to "Mean Red Spider" highlight that he is going to replace his cheating mean red spider with a mean black spider: "I got a mean red spider, and she been webbin' all over town | I'm gonna get me a mean black spider, so I can tear her cobweb down."

In some songs, it is difficult to determine what specific insect the song is about. Examples include "Lazy Bug Blues," "Bug Juice Blues," "Love Bug Blues," and "Biting Bug Blues." But of course, it does not matter; the bug is merely a symbol for a person. Memphis Minnie's partial lyrics to "Biting Bug Blues" reveal that the bug is a metaphor for a woman who is vying for the attention of the singer's man, but the singer exacts her revenge: "Now you girls don't have to worry, because that biting bug is gone | I done stopped that biting bug from breaking up other women's homes."

Later and more modern blues songs with insects and other arthropods as symbols include "Blue Crab Blues," "Scorpion Blues," "Pale White Scorpion Blues," "Spider and the Fly," "Butterfly Blues," "Termite Blues," "Grasshopper Blues," "Drunken Grasshopper Blues," and "This Crab Louse is Killin' Me."

INSECT NAMES FOR EARLY BLUES ARTISTS

Very little is known about Bo Weavil Jackson, but he was a very early country blues musician to be recorded. The guitarist and singer also recorded under the name Sam Butler and all of his eleven known recordings under both names and on the Paramount and Vocalion labels seem to be limited to late 1926. He was probably from the Birmingham, Alabama area.

Bumble Bee Slim's real name was Admiral Amos Easton—an amazing name in its own right. The guitarist and singer was born in Georgia in 1905, traveled all over the US as a musician, and died in Los Angeles in 1968. Unlike Bo Weavil Jackson, Slim recorded more than 150 songs.

INSECTS IN ROCK MUSIC

The blues genre led directly to the genre of rock-and-roll and with it came older songs with insects as symbols. As just one example, "I'm a King Bee," was covered by The Doors, The Rolling Stones, The Grateful Dead, Pink Floyd, Frank Marino, Captain Beefheart and His Magic Band, and many other rock performers.

However, rock music used insects in band names, songs, and album titles without their prior use in blues, folk, or jazz. The examples are too numerous to list for this chapter; see Coelho (2000) for hundreds of examples. Examples of songs released after 2000 include "Dragonfly" by Paul Weller, "Moth" by Audioslave, "Insects" by Finntroll, "All My Friends are Insects" by Weezer, "The World is a Dying Insect" by Kataklysm, "Insect Eyes" by We Are Nots, "Earwig" by Tenement, "Bug" by Wussy, "Mosquito Dance" by Aphid, "7Day Cicada" by LITE, "Hornets! Hornets!" by the Hold Steady, "Wasp Nest" by the National, "Chrysalis" by Opeth, "Termites" by Tall Friend, "Caterpillar" by Mountains of the Moon, "Cocoon" by Barrows, "Locust" by Machine Head, "Crickets" by Drop City Yacht Club, and "Flea vs. Mite" by the Presidents of the United States of America. The progressive metal band Blotted Science released the 2011 album *The Animation of Entomology* with the song "Ingesting Blattaria."

The trickster continues to live on in music. The song "Red Mosquito" by Pearl Jam captures it well: "Watched from the window with a red mosquito | I was bitten ... must have been the devil."

CONCLUSION: INSECTS AS SYMBOLS IN MODERNITY

The modern age. Logic. Reason. Evidence. Scientific method. These hallmarks of the period have arguably eliminated the need for insects to symbolize anything. Profound understandings of cause and effect have revolutionized the age, but insects remain important cultural symbols despite the diminution of myth, supernatural belief, and religion. Insects are symbols of both the cutting-edge of science and technology as well as the environmental backlash caused by both. In addition, as the main topic of this chapter, insects will most likely continue to be used as symbols in popular music to represent many things: metaphors for people, the environment, sustainability, technological backlash, and cultural enemies. Not least of which, insects will continue to serve as tricksters.

CHAPTER SEVEN

Insects in Literature and Language

i see things from the under side now

KARL ZUELKE

In 1916, the newspaper columnist, Don Marquis, ran the first of his series of satirical and comic poems featuring the characters Archy the cockroach and his friend, Mehitabel the cat, in *The New York Sun*. Marquis's column was picked up by *The New York Tribune* in 1922, where it ran until 1930 and from there went into wide syndication in magazine and newspapers around the US until 1936. The column often featured Marquis's other characters, including Freddy the Rat, Warty Bliggins the Toad, and Clem Hawley, better known as "the Old Soak," Marquis's comic alcoholic who struggled through the strictures of Prohibition. But it is through Archy and Mehitabel that Marquis's enduring reputation as a satirist was established.

They were an unlikely duo, Archy claiming to be the reincarnated soul of a "vers libre [free verse] bard," and Mehitabel even more extravagantly claiming to be the reincarnated soul of Cleopatra, residing for her past eight lives in the body of an alley cat. Mehitabel, now on her last life as a feline, swings through the alleys of New York as a carefree Jazz Age flapper cat. "*Toujours gai!*" Mehitabel often sings, and in a poem from a collection of previously lost, unanthologized Archy and Mehitabel poems, discovered in the 1980s in a trunk among Marquis's effects, Mehitabel proclaims her motto: "romance or nothing" (Marquis 1996: 4).

FIGURE 7.1 Archy the cockroach. Archy from an advertisement (1922) in *The New York Tribune*. Courtesy of Library of Congress.

Though Archy and Mehitabel are the best of friends, Mehitabel is nevertheless dangerous to Archy when she's had a bit too much to drink. He is small, and she, after all, is a cat. Archy is imagined to have composed his poems, left in the typewriter in Marquis's office, by launching his body against the keys head first, one at a time—an apt metaphor, often noted, for the struggles of writers everywhere! E.B. White, in his introduction to a 1950 collection of Archy and Mehitabel poems, remarks: "The details of his [Archy's] creative life make him blood brother to writing men. He cast himself with all his force upon a key, head downward. So do we all" (Marquis 1950: xix).

Since he couldn't manage the shift key, the poems are always presented without capitals, and Archy never bothers with punctuation either. Beyond his diminutive physical stature as an insect, which would leave Archy spent on the floor, exhausted once he was done writing, it is Archy's lowly ontological station in this most recent of his succession of lives that so successfully enables Marquis's approach to satire. Archy sees the world from underneath, molding the wit and imagination of a poet and philosopher into a sardonic vision necessitated by his current station as a member of the lowest of the low in the hierarchy of creation, the loathsome cockroach:

expression is the need of my soul
i was once a vers libre bard
but i died and my soul went into the body of a cockroach

it has given me a new outlook on life
i see things from the under side now
thank you for the apple peelings in the wastepaper basket
but your paste is getting so stale i can t eat it
there is a cat here called mehitabel i wish you would have
removed she nearly ate me the other night

(Marquis 2006: 4–5)

Archy is fully representative of the Modern literary insect. He is emblematic of the contradictions inherent in literary and artistic Modernism. Carried over from his past-life history as a poet, Archy embodies the wit, the insight, the perception of an artist. There is something grand in the observations that Archy makes on the events of his day, about the war, Prohibition, labor unrest, the Ku Klux Klan, gangster murders, and other of the pressing issues of the 1920s, as well as subjects of broader and more universal scope such as the transmigration of souls, the value of work versus play, religion, art, and even the ethics of suicide—a topic Archy considers often. What makes Archy Modern is that his view of the political and social state of the world, and of such sweeping philosophical topics as art and religion, are viewed from the "under side," as he terms it. In the end, though he may be the reincarnated soul of a poet, he inhabits his time and place as a bug. Archy combines a sharp and lucid awareness of philosophy, history, literature, and politics with a relentless, sardonic undercutting of the importance of such grandiose conceptions. Archy is a no-BS cockroach, quick to ridicule:

i am kind of leery of
greek gods for the
most part i used to hang around
one of these
greek restaurants a good
deal and all the god i ever
saw there
was the cash register

(Marquis 1996: 85)

Archy is still obsessed, however, with the day-to-day contingencies of the life of a roach. He constantly begs from his "boss" in his poems for more food scraps to be left around the office, and he worries constantly about the physical threats that life as a cockroach present, whether getting inconsiderately squashed, or even eaten by Mehitabel, who being a cat isn't necessarily trustworthy.

Marquis's poem of June 10, 1918, "prohibition rushes toward us" (Marquis 2006: 125–6) provides a good example of Archy's combining of artistic and philosophical awareness and sophistication with a sardonic diminishment of the relevance of that sophistication. Archy tells the story of drinking with some of his cockroach friends in a speakeasy, "making merry with crumbs and | cheese," when "a strange | melancholy descended | upon the jolly crew." Lamenting that soon enough Prohibition will make their communal drinking parties illegal, one of the company of cockroaches notes that, "soon there will | be nothing but that | eheu fugaces stuff," a reference to the Roman poet, Horace, whose line in Latin, "*Eheu fugaces labuntur anni*" translates to, "Alas, the fleeting years slip by." Archy, treating the fancy Latin phrase like a beverage, responds, "it is bitter as wormwood | says i and the | only chaser to it is | the lethean water." The River Lethe, as Michael Sims notes in his annotation of this poem, "flows through Hades, where its waters slake the spiritual thirst of the dead by causing them to forget their lives before" (Marquis 2006: 324). While the "lethean" chaser of their deaths might help them forget their troubles, or perhaps the "lethean water" of alcohol, the roaches are nonetheless alive and aware of their station and their history. They reminisce over their ancestry: "my | ancestors before me have | inhabited yonder ice box," and another claims that, "my ancestors have been | chop house and tavern | roaches for hundreds of years." But, as this roach relates,

> one of my elizabethan
> forbears was plucked from
> a can of ale in the
> mermaid tavern by
> will shakespeare and
> put down kit marlowe s back

Another of the company of roaches relates that,

> i am the last
> of a memorable
> line one of my
> ancestors was found drowned
> in the ink well
> out of which poor
> eddie poe wrote
> the raven

This of course is all pure comic invention, the blending of high culture with roach-level slapstick, but the mixing of high and low, in the Modernist context, is not without meaning.

We can note that an insect characterized at once as both intellectual and vermin has a particularly Modern ring to it, the juxtaposition of opposites being among the characteristic tropes of high Modernism. We see it, for example, in T.S. Eliot's opening lines from his great 1922 poem, "The Wasteland": "April is the cruellest month, breeding | Lilacs out of the dead land, mixing | Memory and desire" (*Norton* 1973: 459), where we have a cruel springing of life from "dead land," as well as the juxtaposition of looking backwards into the past (memory) with a forward-leaning yearning for the future (desire), at the expense of the integrated wholeness of the present.

Archy the Modern poet/roach, the intellectual/vermin, marks a substantial symbolic movement away from an earlier Romantic use of insects as symbols. The Romantics, following the example of the early days of the French Revolution, rejoiced that "the revolution of their epoch would issue in an earthly paradise" (Abrams 1971: 57). This vision of paradise, with potential even greater than the Edenic paradise from which humankind had fallen, follows from early Romanticism's belief in "the theory of general, continuing, and unlimited historical progress" (Abrams 1971: 58). Romantics turned often to nature as a centered and centering symbolic presence, as spiritual and metaphysical guides helping lead the way back to the paradise they believed was coming. Birds were sometimes the creatures invoked for such a purpose, such as in Keats's "Ode to a Nightingale" and Shelly's "To a Skylark," but insects were called upon as well. William Wordsworth wrote two poems titled "To a Butterfly" in 1802. In one, Wordsworth's narrator addresses the butterfly:

> What joy awaits you, when the breeze
> Hath found you out among the trees,
> And calls you forth again!

The narrator's address to the butterfly continues:

> Sit near us on the bough!
> We'll talk of sunshine and of song,
> And summer days, when we were young;
> Sweet childish days, that were as long
> As twenty days are now.
>
> (1984: 254)

We see that the days of childhood, which the presence of the butterfly recalls, were innocent of both memory and desire, so that the relative timelessness of the interaction of innocent child with innocent nature greatly extends the perception of time that still does inevitably pass for the narrator. In the other, the better known of these two poems, Wordsworth also recalls his days as a boy, when he chased butterflies with his sister, but he contrasts the more gentle treatment of "Emmeline" (clearly in reference to Wordsworth's real sister, Dorothy), toward the insects:

Oh! pleasant, pleasant were the days,
The time, when, in our childish plays,
My sister Emmeline and I
Together chased the butterfly!
A very hunter did I rush
Upon the prey: – with leaps and springs
I followed on from brake to bush;
But she, God love her, feared to brush
The dust from off its wings.

(244)

Both of these poems invoke the butterfly as an emblem of the innocence, carefree joy, and wholeness of childhood, all among Romantic literature's most characteristic focuses, and the state toward which they hoped humankind would return.

In America, the Transcendentalist Ralph Waldo Emerson treats an insect—a "humble-bee"—as an embodiment of wholeness as well. The bumblebee was commonly known as a "humble-bee" in Emerson's day, the name perhaps a reflection of the sound made as it flies, but Emerson was aware as well of a pun on the word "humble" in the creature's name. Emerson writes in "The Humble-Bee": "Burly dozing humble-bee! | Where thou art is clime for me ... Wait I prithee, till I come | Within ear-shot of thy hum, – | All without is martyrdom" (1996: 1081–2). Emerson is making a clear statement about the wholeness and natural solace that the presence of the bee's hum provides. Outside of that peaceful, insular realm waits death—"martyrdom."

All beside was unknown waste,
All was picture as he passed.

Wiser far than human seer,
Yellow-breeched philosopher!
Seeing only what is fair,
Sipping only what is sweet,
Thou dost mock at fate and care

Emerson's poem ends with the pronouncement that, "Want and woe which torture us, | Thy sleep makes ridiculous." The bee's centering solace functions as a corrective, the contrasting alternative to the waste, martyrdom, and ridiculousness of the world beyond the reach of his hum. "The Humble-Bee" thus frames a characteristic Transcendentalist statement, from one of the guiding lights of the movement, who believed that humankind could turn to nature and see through it for a glimpse of the divine presence that inspirits all of creation.

A brief discussion of Romanticism and insects matters in a treatment of Modernism and insects because Romanticism, through its desire to elevate and celebrate the centering presence of nature, can nevertheless be regarded as an attempted defense against the effect of forces that would eventually lead to the Modern predicament, the inevitable sliding away of a human relationship with any kind of philosophical focus. This includes nature, but it also includes concepts such as God and the primacy of religion, patriotism, and even progress. The French Revolution had degenerated into a blood-thirsty nightmare. The burgeoning of industry, with the spread of its "dark Satanic Mills" of William Blake's famous phrase, was impossible to ignore and didn't at all look to be receding in favor of heaven in England, as Blake had hoped. This disintegrating slide away from a unifying and centering Romantic notion of progress toward heaven on earth, and a consequent grasping at or a yearning for what was being lost, is expressed repeatedly in Modernist literature, and this perhaps paradoxically occurs along with a simultaneous hastening of a withdrawal from the center through acts of deconstruction, re-envisioning, or even vandalism. A full century after the Romantics, William Butler Yeats, in "The Second Coming," declares flat out, for example, that, "Things fall apart; the center cannot hold; | Mere anarchy is loosed upon the world" (1986: 89); T.S. Eliot, in "The Love Song of J. Alfred Prufrock," expresses the idea as aging, the separation of the narrator's psyche from its youth and vitality: "I grow old ... I grow old ... | I shall wear the bottoms of my trousers rolled" (*Norton* 1973: 452); Wallace Stevens, in "The Emperor of Ice Cream," understands that everything is *melting*: "The only emperor is the emperor of ice-cream" (*Norton* 1973: 248); Matthew Arnold, in an earlier, nineteenth-century Modernist statement, draws an analogous idea in "Dover Beach" regarding the primacy of religious faith:

> The Sea of Faith
> Was once, too, at the full, and round earth's shore
> Lay like the folds of a bright girdle furled.
> But now I only hear
> Its melancholy, long, withdrawing roar,
> Retreating, to the breath
> Of the night-wind, down the vast edges drear
> And naked shingles of the world.

(1994: 86)

Ernest Hemingway's short story "A Clean, Well-Lighted Place" lays out the same idea strongly and clearly:

> It was all a nothing and a man was a nothing too. It was only that and light was all it needed and a certain cleanness and order. Some lived in it and never felt it but he knew it all was nada y pues nada y nada y pues nada. Our nada who art in nada, nada be thy name thy kingdom nada thy will be nada in nada as it is in nada. Give us this nada our daily nada and nada us our nada as we nada our nadas and nada us not into nada but deliver us from nada; pues nada. Hail nothing full of nothing, nothing is with thee.
>
> (1925: 383)

And the philosopher Friedrich Nietzsche, in *Thus Spake Zarasthustra*, proclaims famously that "God is dead" (1927: 6), by which he means that the possibility even of belief in God has become inescapably problematic in humanity's Modernist conception of its place in the cosmos.

Sure, but what about *insects*? Robert Frost's sonnet, "Design," published in 1922, stands as one of the preeminent expressions of high Modernism. It involves a spider and a moth. The poem is presented here in its entirety:

> I found a dimpled spider, fat and white,
> On a white heal-all, holding up a moth
> Like a white piece of rigid satin cloth –
> Assorted characters of death and blight
> Mixed ready to begin the morning right,
> Like the ingredients of a witches' broth –
> A snow-drop spider, a flower like a froth,
> And dead wings carried like a paper kite.
>
> What had that flower to do with being white,
> The wayside blue and innocent heal-all?
> What brought the kindred spider to that height,
> Then steered the white moth thither in the night?
> What but design of darkness to appall? –
> If design govern in a thing so small.
>
> (Various 1922: 38)

Frost's poem begins with the description of a scene in nature, the commonplace occurrence of what is likely a white crab spider (a spider which does not spin a web) on a white flower, and which has captured and is eating a white moth. The description is not neutral or merely observational, however, as the creatures

FIGURE 7.2 "I found a dimpled spider, fat and white." Photograph by Jessee J. Smith.

are represented as "characters of death and blight," a decidedly judgmental characterization of the spider and the insect. The creatures are given further negative significance when their presence is compared to the ingredients of a "witches' broth," suggesting an accusation of malevolence and evil. The scene is set, then, for a series of questions that follow in the second stanza of the poem: Why was the flower white, like the white spider and the white moth, when the heal-all (or self-heal) flower is generally blue? The framing of the question, innocuous enough on its surface, nevertheless implies that something deeper is at work, which is brought to the surface in the next question: How was it that the moth and the spider were brought together? Is there meaning here? This is answered, in effect, by the third question, which is explicit in its accusation: What but design of darkness to appall? Some sort of infernal design must be at work. This question-as-answer confirms the implication of the description of the scene from the first stanza, where death, blight, and a witches' broth are all brought to bear on what might seem an otherwise commonplace instance of predation in nature. Nature, in this conception, is not benign, and it is certainly not an example of the innocent delight and wholesomeness that Wordsworth invokes with his butterflies and Emerson invokes with his humble-bee. Nature is frankly determined to be representative of an overarching or fundamentally inspiriting evil. Rosanna Warren maintains that we see in "Design" "a harrowing meditation on the idea of design in creation, and on the human appetite for order" (2015: 30). The design in creation is harrowing because, as Frost is implying, it is ordered on the basis of an inspiriting malevolence. The moth is surely innocent, and its demise at the hands of a predator is a function of a "design of darkness to appall." Deirdre Fagan and Robert Seltzer note that, "Many critics have interpreted 'Design' as an example of Frost's firm belief in divine design" (2010: 50). This divine design in Frost's conception of the poem, however, whether evil or benevolent, would seem to run against the Modernist grain as I have characterized it. A malevolent design in nature is hardly as comforting as Emerson's humble-bee, buzzing through the flowers in the garden and sanctifying everything touched by its hum, but it nevertheless is reference to a world inspirited by something, a design, and by extension, a designer. The spirit is malevolent, but it is spirit. The cosmos is shot through with meaning, a conception that runs against the Modernist falling away from an inspiriting order. But this reading ignores the effect of the poem's final line. Warren as well claims that the final line introduces something quite different to the poem:

> "Design" ... presented a hypothesis about the authorship of the design—some dark will, some cosmic malice. The last line withdraws that perhaps infantile suggestion, leaving us with the bleaker notion that the universe makes no moral sense, malicious or beneficent. "If design govern in a thing so small." The tone is crucial in bringing off the effect: unpretentious, semi-casual, sardonic, not a wisecrack, but a dismissal.
>
> (2015: 30)

The implication of some cosmic design, perhaps at the hands of God the Creator, or at the instigation of something darker, is thus dismissed by Frost in the poem's final wondering about whether design is truly apparent in such a small, seemingly insignificant occurrence. The very act of stating that supposition implies that there really may not be a designer, whose work brings the moth and spider together. The poem's title, then, is ironic. Everett Carter takes it a step past irony in his estimation of Frost's purpose. "It is more likely a poem whose principal tone is mockery" (Carter 1988: 24). Frost is mocking, in Carter's estimation, any notion of the existence of God or the devil, as well as the belief of poets like Wordsworth and Emerson, "sentimentalists who read human emotions into the world of lower organisms" (Carter 1988: 24).

Thus the spider, feeding on the unlucky moth, finally emerges as a Modernist symbol of something grimmer, even more unsettling than the presence of a malevolent designer. It signifies nothing. Creation is uncentered, uninspirited. An existential void yawns, and the door to nihilism can now be imagined.

Frost's spider as existential symbol is a powerful statement of Modernism's abandonment of the recognition of the divine in creation, a belief that had sustained much of western philosophy for millennia. But the image, while local, achieved as a result of close and intimate observation, almost as if through a magnifying glass, is nevertheless somewhat distanced because of its non-human impersonality and its separation from a human frame of reference. This insect and this spider are something other, something alien, so the existential impact of their predator/prey association remains at a distance from the human observer. Are there any instances of insect imagery and symbolism that close the perceptual distance between the insect and the human frames of reference? Franz Kafka's novella, *The Metamorphosis*, does exactly this, and in its horror, the novella is celebrated as one of Modernism's defining statements.

It is perhaps worth noting that the closing and indeed the merging of the human and the insect frames of reference doesn't necessarily invoke horror. The image of the fairy, for example, the diminutive humanoid chimera of human and insect, is often conceived as a figure of delight, something the Disney Studios take advantage of in their characterization of the beautiful and sexy fairy, Tinkerbell. Jiminy Cricket—another Disney character, the upstanding conscience of the boy/puppet Pinocchio—is a cricket, though with his top hat and umbrella his thorough education in the ethics of good behavior, his angelic singing voice, and his two arms and two legs, he comes off more as a diminutive human rather than anything resembling an actual cricket. Archy is an effectively congenial cockroach because his soul is human. More often, though, the juxtaposition of human and insect proves disturbing. Hollywood's and literature's most frightening characters often combine the human and the animal—vampires are part bat, werewolves are part wolf, the Creature from the Black Lagoon is part fish. H.G. Wells's *The Island of Dr. Moreau* is inhabited by a freakish contingent of human/animal hybrids. Hollywood's *The Fly* combines a human with the small, excrement-eating creature humans routinely swat in annoyance and disgust to a particularly horrifying effect. This is the level of horror that *The Metamorphosis* opens with.

The opening line is among the most well-known in the world's literary canon because of its shocking establishment of the nightmare that is to follow: "As Gregor Samsa awoke one morning from uneasy dreams he found himself transformed in his bed into a gigantic insect" (Kafka 1961: 67). While often mistakenly conceived of by readers as a cockroach, the details of Gregor's new body, with its hard shell, mark him as a beetle, most probably a scarab or a dung beetle.

That Gregor emerges from his dreams as a dung beetle is considered by some critics to be significant. There is little in western culture that conceives of the cockroach as anything beyond verminous and repugnant. Because of the psychology of Gregor's character, often interpreted as an autobiographical invention, reflective of Kafka's anxiety and self-loathing, it is tempting to read Gregor immediately as the lowest of the low as well. Lewis W. Leadbeater argues, however, that the dung beetle, "Especially as it appears in classical literature, has always been a creature of two dimensions. That is, the gross, ugly, and repulsive side of its physical nature is usually counterbalanced by a quasi-divine association" (1986: 170). This matters in the novella because part of Kafka's strategy is to maintain the reader's empathy for the struggles of Gregor, at least at first. The reader learns that he is hard-working, is supporting his family, and even managing to put extra money away so that in five or six years he will be able to alleviate the debts his father amassed when his business failed. While he dreams of eventually leaving his job as a commercial traveler, constantly catching trains and conducting the business of the company in distant locales, he is nevertheless a trustworthy and loyal employee. But when he misses his morning train and is inexplicably absent from the office, his supervisor arrives at the family flat to find out what is wrong. "What a fate," thinks Gregor, "to be condemned to work for a firm where the smallest omission at once gave rise to the gravest suspicion!" (Kafka 1961: 74). Gregor is clearly caught in the plight of the underappreciated worker, and while he wishes to quit, his responsibilities prevent him from leaving the firm. Rather, he reflects bitterly, "Was there not among them one single loyal devoted man who, had he wasted only an hour or so of the firm's time in the morning, was so tormented by conscience as to be driven out of his mind and actually incapable of leaving his bed?" (74).

While a bitter reflection on Gregor's part, it is actually comic. While Gregor is still occupied with worries about catching his morning train, and is so guilty about it that he indeed feels incapable of leaving his bed, the reader is fully aware that Gregor doesn't leave his bed because he can't control his new insect body well enough to manage getting out of his bed. He is lying on his back, waving his six pitifully skinny legs in the air, trying to work out a way to reach the floor and ultimately the door. Gregor is trapped in his new nightmare beetle's body.

The first section of the story is concerned with the physical details of his insect body and his futile attempts at trying to control it. Gregor's struggles recall Archy the cockroach in one respect: his is a fully human consciousness residing in an insect's body, and while the result is comedy, the comedy is of an entirely different order from Archy's. Archy is in control of himself. It is the vast distance between the realms of human and insect—which is nevertheless fully contained within Archy—that enables his wry observation of the human political and social fabric. In fact, it is Archy's control of his body that permits him the possibility of communication in the first place. While flinging his body headfirst against the typewriter keys beats him up and exhausts him, he manages the act of writing with a measure of skill. Gregor's comic struggle

FIGURE 7.3 *Carabus granulatus*. Photograph courtesy of Niels P. Nielsen/Getty Images.

arises from his utter unfamiliarity with his terrible predicament, and the comedy is a nightmarish physical comedy. He can't even get out of bed at first. His human aspect is tormented by the consequences he fears his absence at work will trigger. But the reader is engaged with the closely observed details of his feet, his position on the bed, his fear of bumping his head against the dresser, the way he finally manages to turn the key in its lock on his door:

> Then he set himself to turning the key in the lock with his mouth. It seemed, unhappily, that he hadn't really any teeth—what could he grip the key with? – but on the other hand his jaws were certainly very strong; with their help he did manage to set the key in motion, heedless of the fact that he was undoubtedly damaging them somewhere, since a brown fluid issued from his mouth, flowed over the key and dripped on the floor.
>
> (80)

The comic horror of the scene is generated through the disjunction of Gregor's mind and spirit from the workings of his insect body. The distance between them is minimal, yet the disjunction is absolute. Gregor can't work his strange body because his mind is still fully human, with all its attendant guilt and anxiety intact. The sense of revulsion so striking in the scene comes partly as a result of the revulsion many people experience at the sight of certain insects. If they're fuzzy and they hum, like a bee, or they're soft and colorful, like a butterfly, that revulsion may be minimized. Other insects, such as roaches, grubs, etc., lend themselves much more quickly to horror when contemplated too closely. This comes because insect morphology is utterly alien to most human sensibilities. But it is the disjunction between the realms of insect and human, positioned nevertheless within the same chimeric creature, that is just as responsible for the reader's revulsion. Kafka focuses on the alienating details of insect bodies to maximize the horror generated by Gregor's strange predicament, the segmented belly, the skinny waving collection of legs, the odd jaws dripping brown fluids, the painfully sensitive area at the end of Gregor's abdomen, the sticky substance on Gregor's feet that leaves wet spots on the walls and ceiling as he crawls over them. It takes a series of experiments and clumsy, painful mistakes for Gregor to figure out how it all works, and in the meantime, his sister, mother, father, and boss are right outside the door, knocking and calling to him, frantically trying to determine why he hasn't gotten up, opened the door, and caught his train.

Frederick Karl maintains that

> Kafka achieves here the quality of all horror stories. While one line of development indicates a terrible change, another, here the father, proceeds as if nothing has happened. Only the reader is aware of both levels and can, from the start, respond to them. Kafka has so decentered or deconstructed the typical family or office story that its components leave the reader nowhere to settle. So too Gregor. Even as his new body is a fact, he looks forward to the moment when his morning's delusion will fall away, going as far as attempting to carry out his usual ablutions. What makes his condition so intense is that even as he hates every moment of his earlier life, all he wants to do is return to it.
>
> (1991: 466)

As bad as his former life was, with all its anxieties, pressure, thanklessness, and boredom, it was still better than the predicament he finds himself in. Yet the transformation of Gregor seems to be a symbolic manifestation of the abysmal worth of his former life. His former opinion of himself, in this reading of his metamorphosis, was so low that he was essentially little more than an insect in his and everyone else's esteem anyway. He has simply transformed one evening into what he and everyone else saw him as in the first place. But as Leadbeater's quote about dung beetles above hints, Gregor's transformation into a dung beetle, while at once disturbing and disgusting to himself and the members of his family, also hints at other manifestations of his transformation. On one hand, Gregor may be asserting himself against the ungrateful advantage that his family has taken of his work. Karl suggests as much when he asks, "Is the alteration of body, and eventually of mind, a wish fulfillment, possibly a means of gaining revenge on a stressful and ungrateful family?" (1991: 466). Beyond self-image as vermin, and as possible revenge, many critics even read Gregor's transformation into the quasi-divine scarab as making him into something of a Christlike figure in that the sacrifice he makes of his body, eventually his life, provides for the salvation of his family. No longer able to depend on Gregor's support, his father is forced back into the world and finds strength and success again: "The man who used to lie wearily sunk in bed whenever Gregor set out on a business journey ... Now he was standing there in fine shape; dressed in a smart blue uniform with gold buttons ... from under his bushy eyebrows his black eyes darted fresh and penetrating glances" (Kafka 1961: 108). When

Gregor dares to show himself to the family's lodgers, entranced as he was by his sister's violin playing, his father drives him back into his room with a salvo of apples, one of which causes a wound that eventually proves fatal to Gregor. This assertion of himself against the horrible aspect of the giant beetle living in his son's room is something unprecedented for Herr Samsa. Gregor's sister, as well, begins taking classes and finding ways to better herself in society.

If Gregor's metamorphosis amounts to a sacrifice which leads to the betterment of his family, we can certainly see parallels with the crucifixion story of Christ's life. In Christian mythology, out of the most extreme depths of humiliation and torment that must characterize the crucifixion experience arises the promise of a glorious future and salvation for all humankind. While certainly less sweeping than the Christian story, the parallel is clear: out of the extreme humiliation, pain, disgust, and eventual death of Gregor as an insect, the turn of his family's fortunes is provided for. As Aichele and Walsh maintain, "Kafka's work is fantastic [i.e. a fantasy]. Alongside such works, the gospel transfigurations and their recklessness with bodies bespeak the aura of horror. The Gospel speaks of this vanishing act, this transfiguration, as a glorification/deification" (2011: 273).

The Christian story historically resolved into a glorification of Christ's death that is recognized and celebrated by the faithful, and in this respect, comparisons with *The Metamorphosis* eventually falter. Gregor's family responds to his change at first with understandable confusion and horror, though his sister, Grete, takes care of him as well as she can, providing him with a variety of foods and trying to communicate with him in ways the rest of the family can't bear. But eventually, even she understands that Gregor can no longer be seen as a member of the family:

> "He must go," cried Gregor's sister. "That's the only solution, Father. You must just try to get rid of the idea that this is Gregor. The fact that we've believed it for so long is the root of all our trouble. If this were Gregor, he would have realized long ago that human beings can't live with such a creature, and he'd have gone away of his own accord. Then we wouldn't have any brother, but we'd be able to go on living and keep his memory in honor."
>
> (Kafka 1961: 125)

Eventually Gregor dies after his sister locks him for one last time in his room, emptied now of furniture except for a bureau he hides under, and filled with decaying bits of uneaten food and a thick layer of dust. When the charwoman finds his body in the morning, she simply disposes of it, and the family doesn't even wish to hear the story. The family decides to move away, since the apartment had been chosen by Gregor anyway. Riding on a train out of town together for an excursion into the country, now that they're finally free of the overbearing and burdensome presence of the beetle in Gregor's old room,

> Leaning comfortably back in their seats they canvassed their prospects for the future, and it appeared on closer inspection that these were not at all bad, for the jobs they had got, which so far they had never really discussed with each other, were all three admirable and likely to lead to better things later on.
>
> (132)

Freed from the loathsome insect, they are finally able to pursue their lives.

Much of the criticism regarding *The Metamorphosis* concerns the autobiographical aspect of Gregor, with, as Leadbeater explains, "Special emphasis on the sociological and psychological ramifications of the novella as they pertain to the disintegration and revitalization of the Samsa family through the transformation and eventual disappearance of Gregor Samsa" (1986: 169). Disintegration, as Suzanne Wokenfeld explains, is central to all of Kafka's work:

> The misunderstandings, the confusion, the despair that permeate Kafka's writings derive from a vision of disintegration. Although the symbolic images in which Kafka expresses this vision allude, often simultaneously, to man's relation to himself, to society, and to God, it is the psychological perspective that provides the core from which the sociological and theological implications are refracted. The rupture is within.
>
> (1976: 25)

Thus, the disintegration of Gregor is often read as emblematic of the disintegration that Kafka himself felt in the face of his difficult relationship with his father. "Through his writing Kafka was able to shape and externalize the morbid feelings of guilt, shame, fear, and impotence that characterized his relationship with his father" (Wokenfeld 1976: 25). In spite of the pitiful end that Gregor eventually succumbs to, however, Frederick Karl insists that, along with several other of Kafka's stories,

> It is altogether incorrect to read these stories, from "The Metamorphosis" on, as defeatism or withdrawal into nonentity status. Kafka has in fact located the enemy in that post-Freudian world, and the enemy is anxiety, uncertainty, fear. … Kafka's so-called defeatism rarely involves meek surrender; quite differently, he insists on challenging the specters on the very ground he has laid out, and while he may lose, the battle is, apparently, the main thing.
>
> (1991: 736)

FIGURE 7.4 Franz Kafka, 1923. WikiMedia Commons, https://commons.wikimedia.org/wiki/File:Franz_Kafka,_1923.jpg.

So, in *The Metamorphosis*, Kafka took internal feelings of personal disintegration, anxiety, etc., and in portraying them in such disturbing and brilliant fashion, projected them onto the world's canvas and created one of the defining statements of Modernist disintegration. The personal, through art, with Kafka's great dung beetle as a vector, becomes universal. We identify with Gregor as insect, because we participate as well in the Modern (now post-Modern) world through our own disjunct and disintegrated anxieties as members of a disjunct and disintegrating society. In its parts, which we can find within ourselves, we discern the fragmenting whole.

If, as Karl maintains, Kafka's stories, *The Metamorphosis* in particular, can be read as challenges to the Modernist specters of anxiety and fear of loss—which Kafka felt personally, to an intense degree and challenged as a heroic artist—the challenge also has a self-destructive quality. Karl notes as well that, "It is a premise of Modernism that each avante-garde is a kind of suicidal charge. Its thrust is a kamikaze attack in which both the art and the artist are burned up or burned out" (1991: 135).

Beyond even art and the artist, Karl maintains that,

> The end of the century, despite its progressivism, had become a culture of death. The avant-garde as the avatar of revolution or civil war turned the arts into elements that had to die or self-destruct in order to make way for the new, whatever its characteristics .... For Kafka, there was the sense of

> an ending, his perception that whatever had existed no longer functioned sufficiently for adequate human life.
>
> (138)

The "culture of death," in which Kafka was immersed in 1912 when he wrote *The Metamorphosis*, of course came to full fruition just a few short years later in the First World War, in which millions of soldiers and civilians met their deaths in the unprecedented conflagration that left Europe physically, politically, and spiritually devastated. While Kafka didn't commit suicide—he died at age forty-one of tuberculosis in an Austrian sanitarium in 1924—he attempted a kind of artistic suicide. He is thought to have burned over 90 percent of his work, and in a letter to his friend, Max Brod, his death approaching, Kafka wrote, "My last request: Everything I leave behind me ... in the way of diaries, manuscripts, letters (my own and others'), sketches and so on, to be burned unread" (Batuman 2010). Brod disregarded his friend's request.

The first half of the twentieth century was a dark time, so it is understandable that its art would reflect the anxieties of the age. Even Don Marquis, remembered still as one of America's pre-eminent humorists, left a legacy through Archy and Mehitabel that is reflective of the anxieties of his times, sometimes resigned to them, sometimes satirizing in fury over them. Mehitabel may have proclaimed herself *toujours gai*, but she lived homeless in an alley and often went hungry. "Romance or nothing," her motto, resulted too often in nothing. Archy, fond of a jovial round of drinks with his rat, toad, and cockroach companions, might just as frequently sink into depression and satiric bitterness, reflective of Marquis's complicated artistic ambitions and his ambivalent take on the state of his country and the world. Satire, as Michael Sims notes, is often born in outrage (Marquis 2006: xv). Sims makes the point that satire is often categorized as being modeled off of one of the two great Roman satirists, Horace and Juvenal. Horace "comes across in his satires as mildly amused at his fellow human beings, shaking his head in a will-they-ever-learn sort of way" (xv). Juvenal, on the other hand, "is distinctly not amused" (xv). Mark Twain, the greatest of America's satirists, felt himself to be squarely a Juvenalian in his approach to social commentary. Sims quotes Twain's letter to William Dean Howells: "I don't ever seem to be in a good enough humor with ANYthing to satirize it; no, I want to stand up before it and curse it, and foam at the mouth – or take a club and pound it to rags and pulp" (qtd. in Kaplan 1966: 169). Sims claims that, "Don Marquis often felt the same way" (2006: xv).

In "the cockroach who had been to hell" (Marquis 1961: 31–2), Archy writes about an acquaintance, "an old simp cockroach," who thinks he has been to hell and who describes his experience with flames, winged demons and sulfur. Archy tells his friend, "what you | beheld was not hell all that | was natural someone was fumigating | a room and you blundered | into it." Archy then relates his story of the transmigration of his soul from vers libre poet to his present cockroach incarnation. The poem up to this point is more of Marquis's cockroach slapstick, until, "you re | irreligious says the old simp," who with the other roaches of the gathering cry "atheist" and advance on Archy with the intent of murdering him for his lack of religious belief. The comic poem has taken a sudden dark and violent turn, where a reader can sense how appalled Marquis himself must be at the murderous intolerance of religious fundamentalism, an all-too-common historical occurrence. Fortunately, Archy is saved by the appearance of a friend. He slips back into the mode of the world-weary philosopher:

had it not been for freddy
the rat i would now be
on my way once more i mean
killed as a cockroach and transmigrating
into something else well
that old whitebearded devil is
laying for me with his
gang he is jealous
because i took his glory away
from him don t ever tell me
insects are any more liberal
than humans

While Archy's mood can change quickly, "world weary" describes his overall emotional state. It's not surprising that this reflects Marquis's emotional state as well. With Marquis harboring an ambition to be taken seriously as a writer, he admitted that, "I got to see my column as a grave, twenty-three inches long, into which I buried myself every day" (Marquis 2006: xxx–xxxi).

He retired from the column full time in 1925 to pursue his other work, including a somber crucifixion drama that didn't see much success, and a comedy play that ran on Broadway featuring Clem Hawley, the "Old Soak," that was quite popular. Marquis played Clem himself in some of the productions. His last years were tragic, though, as he lost his first wife and both of their children, and his second wife as well. Grief-stricken, Marquis suffered a series of strokes and died nearly paralyzed in 1937. His moods and even his depression, as these difficult circumstances forced their way into his life, found expression through Archy, who repeatedly contemplates suicide. In Marquis's poem, "the wail of archy," the cockroach makes one of his suicidal attempts: "i with the tastes of a

FIGURE 7.5 Don Marquis. George Grantham Bain collection, Library of Congress. Wikimedia Commons, https://commons.wikimedia.org/wiki/File:Don_Marquis.jpg.

byron | expected to live upon garbage … i with the soul of a hamlet | doomed always to wallow in farce." Archy, in full-throated wail, goes on:

> yesterday maddened with sorrow
> i leapt from the woolworth tower
> in an effort to dash out my brains
> gods what a wretched pathetic
> and anti climactic attempt
> i fluttered i floated i drifted
> i landed as light as a feather
> on top of a bald man's head

(Marquis 1961: 82)

While the scene is undoubtedly funny, the frustrations and full-throated wailing of the poet, of Marquis himself, can be discerned beneath the farce.

FIGURE 7.6 Archy at work. Courtesy of Internet Archive, https://archive.org/details/livestimesofarch0000marq/page/n3/mode/2up.

The tribulations of a troubled age found expression through insects as symbols in Frost, Kafka, and Marquis. There are other examples in the literature of Modernism, of course—the fleas behind the fictional Algerian plague that drive Albert Camus's novel, *The Plague*; the lice in Arthur Rimbaud's poem, "The Lice Hunters," written in 1871, which can still shock readers through its annihilation of gentile poetic decorum, opening with the image of a lice-infested boy with open red sores on his face upon whom two nuns are, with "royal nails electrically mashing | to crackling death the gray and lazy lice." The odd scene gives rise to the moment of the boy's sexual awakening, leading to a classic Modernist statement: "and the child feels, under the slow caresses, | welling and sinking an endless need to cry" (*French* 1958: 79). Vladimir Nabokov, the world-famous Russian writer who emigrated to America and wrote brilliantly in his new, adopted English, including such works as *Lolita*, *Pale Fire*, and other highly acclaimed novels and short stories, also pursued an equally illustrious career as a lepidopterist, making important scientific contributions to the study of butterflies. Butterflies and the phenomenon of metamorphosis arise frequently as symbols in Nabokov's fiction. David Larmour maintains that,

> The icon of the butterfly is suggestive of the author's transformation from a Russian into an American writer .... In the Nabokovian cosmotext,

> emergence into the butterfly from previous stages of existence becomes a metaphor for general ideas of transformation, rebirth, immortality, and contact with the world hereafter, and, more specifically, for the emergence of the writer into a creative artist and the adaptation of the émigré to his new life and language.
>
> (2000: 62)

There are more recent examples of literature invoking insects as symbols as well. Gabriel Garcia Marquez's novel *One-Hundred Years of Solitude* features a scene where one character is followed everywhere he goes by a cloud of yellow butterflies. William T. Vollman's 1987 novel, *You Bright and Risen Angels*, presents an intricate, fantastical parable about a war between the inventors of electricity and the insects. More recently Sue Monk Kidd's 2002 coming of age novel, *The Secret Life of Bees*, features a main character who runs away from her abusive father and takes up a new life as an apprentice beekeeper.

The Secret Life of Bees opens with an epigraph from *Man and Insects*, an observation that bees will quickly sense that their queen has been removed from the hive and within hours will show the behavioral signs of queenlessness. This helps sets up a story about women who support and depend on one another. Bees enter the bedroom of the protagonist, Lily, who watches them buzzing around her room at night. Lily remarks that, "I want to say that they showed up like the Angel Gabriel appearing to the Virgin Mary, setting events in motion I could never have guessed" (Kidd 2002: 2). The family's housekeeper, Rosaleen, remarks that, "Bees swarm before death" (3), another symbolic use of the insects, forewarning of the challenging circumstances for Lily and Rosaleen that are soon to arise. The symbolic presence of the bees is the thing, acting as symbol of the unity of a community of women (both queen and worker bees are female), and as a portent of upcoming difficulties.

Often in the literary evocation of insects, they function as symbols—this is certainly the case with Frost's "Design"—or they act as stand-ins for the author, grotesque or satiric/comic masks through which anxiety-ridden or frustrated writers manage to project their personas. This is perhaps not surprising given the vast differences between the human realm and the world of insects, the alien character of insects, their small size, their bizarre physical structure, their lack of anything like a recognizable intelligence, their behaviors seemingly driven so relentlessly by blind instinct. Literature abounds with dogs, horses, cats, and numerous other higher-order animals whose intelligence and familiarity lend themselves to an easy anthropomorphism, but outside of the genial and empathetic spider of E.B. White's *Charlotte's Web*, literary arthropods tend to stand for something other, something larger than themselves. In some ways this is to be expected, since symbol-making is one of literature's most effective means of contributing to human self-understanding, but it does little to help us understand insects as insects. For writers who attempt this we need to turn to the essay and non-fiction literary forms, and to nature writing, but even then, the temptation to turn insects into symbols can arise.

J. Henri Fabre, the French writer and amateur entomologist, spent much of his life observing insects. He did his work in the latter half of the nineteenth century, living in poverty, pursuing his observations and experiments with only the most rudimentary equipment. Fabre didn't even own a microscope, depending instead on a small, inexpensive hand lens for many of his observations. He wrote children's science books, taught science at a *lycée* near Avignon for twenty years, and despite his poverty and obscurity, was able to build a reputation among scientists for the careful and detailed observations he made of insect behavior. Just before his death, in the early years of the twentieth century, Fabre's work finally came to the attention of the wider world and made him famous. Charles Darwin himself was aware of Fabre's work early on, before this widespread recognition, calling him an "incomparable observer" (Fabre 1991: xvi), though as Edwin Way Teale notes, "although Fabre had the highest personal regard for the English scientist, he was so little interested in *The Origin of Species* that he never read more than a few pages. He was a realist opposed to hypotheses. His harvest was a harvest of facts rather than conclusions" (Fabre 1991: xvi).

Fabre did the bulk of his work during an age of natural history fact-gathering, those masses of accumulating facts furnishing the stockpile of raw natural history material out of which theorists like Darwin and Alfred Russell Wallace (who were both inveterate natural history fact-gatherers themselves) eventually developed their monumental theories about natural selection, evolution, and the very existence of the astonishing diversity of life on Earth. What made Fabre's work unique, as Teale observes, is that he "led in the study of *living entomology* at a time when the science seemed preempted by those whose horizon of interest was limited to the dead insect and the pinned specimen" (Fabre 1991: xvii).

Despite his reputation as an incomparable observer, however, Fabre's language can often take on metaphoric import, perhaps almost in spite of himself. Describing a wasp of the genus *Sphex*, Fabre writes,

> The Sphex has shown us how infallibly and with what transcendental art she acts when guided by the unconscious inspiration of her instinct; she is now going to show us how poor she is in resource, how limited in intelligence, how illogical even, in circumstances outside of her regular routine. By a strange inconsistency, characteristic of the instinctive faculties, profound wisdom is accompanied by an ignorance no less profound.
>
> (1991: 55)

Fabre's dismissal of the insect's intelligence would likely not elicit an argument from most casual observers, though as Teale claims, entomologists working in America "demonstrated that instinct provides considerable latitude in the actions of many insects, a finding that has been substantiated by numerous more recent investigations" (Fabre 1991: xvi). For Fabre, instinct guides behavior in the insect like rails guide a train, laying out a predetermined path from which deviation is impossible. He gives the example of a digger wasp of the genus *Philanthi*, the females of which will sting and carry away a foraging honey bee to provide food for her larva. If the female wasp senses that the bee has a crop filled with nectar, she will squeeze the bee before she lays an egg in it to feast on the disgorged nectar. Fabre observed a praying mantis in turn capture a wasp while it was eating its captured bee's nectar:

> While the Mantis held her transfixed under the points of the double saw and was already munching her belly, the Wasp continued to lick the honey of her Bee, unable to relinquish the delicious food even amid the terrors of death. Let us hasten to cast a veil over these horrors.
>
> (Fabre 1991: 57)

In an experiment, Fabre observes that the *Sphex* wasp uses the antennae of the paralyzed bee as handles to transport it to her burrow. Fabre cuts the antennae off of the paralyzed bee. The wasp, who could just as easily grab a leg and carry away the bee, cannot figure out how to do it, and thus abandons the prey, unable to manage even the simplest adjustment of its absolute and rigid behavioral instinct, which is minutely almost miraculously adjusted to the prey bee's body but which is also incapable of adaptation to changes in it.

Fabre seems to be discerning a two-fold phenomenon regarding insect behavior. On one hand, their behavior is machine-like. The wasp is incapable of altering its behavior in the face of circumstances for which its instinct leaves it unprepared. On the other hand, there is both profound wisdom and horror in insect behavior. Their actions appear minutely clever and even amazing, signifying an overarching wisdom. So whether wise or horrible, the insect is unconsciously acting out the intentions of something far beyond its immediate circumstances. Fabre, uninterested in a theory of natural selection and evolution which could explain the development of such behavior, rather falls back on a conventional assumption without articulating it as such: the insect was designed by, and in its behavior reveals the presence of, something greater, probably a wise or malevolent spirit animating the creation of which it is an element. Not yet a Darwinist, and focused on detail and experiment, Fabre betrays a fundamental religious streak. The wasp is showing us something greater than itself. Fabre is no Modernist, in other words!

A full century after Fabre, Annie Dillard, in *Pilgrim at Tinker Creek*, observes examples of the same inflexible patterns of insect behavior, and she leans on Fabre's observations as well, but reaches a Modernist conclusion about its significance quite different from his. Dillard muses that, "Fish gotta swim and bird gotta fly; insects, it seems, gotta do one horrible thing after another" (Dillard 1974: 63). She follows Fabre's description of the mating of the praying mantis:

> The mating rites of mantises are well known: a chemical produced in the head of the male insect says, in effect, "No, don't go near her, you fool, she'll eat you alive." At the same time a chemical in his abdomen says, "Yes, by all means, now and forever yes."
>
> While the male is making up what passes for his mind the female tips the balance in her favor by eating his head. He mounts her. Fabre describes the mating, which sometimes lasts six hours, as follows: "The male, absorbed in the performance of his vital functions, holds the female in a tight embrace. But the wretch has no head; he has no neck; he has hardly a body. The other, with her muzzle turned over her shoulder continues very placidly to gnaw what remains of the gentle swain. And, all the time, that masculine stump, holding on firmly, goes on with the business! ... I have seen it done with my own eyes and have not yet recovered from my astonishment.
>
> (57–8)

Dillard follows another of Fabre's observations, about caterpillars called pine processionaries, which have the habit of laying down silk trails for others to follow in a procession. Fabre guides them onto the circular rim of a large vase and watches in astonishment as they process around the rim for seven days (65–7). Dillard remarks, "That the insects have adapted is obvious. Their failures to adapt, however, are dazzling. It is hard to believe that nature is partial to such dimwittedness" (65). In observing nature on her own—the striking and horrifying spectacle of a predatory giant water bug sucking the juices from a frog haunts her for years after witnessing it—and comparing her observations about nature's seeming obliviousness to the value of an individual, which is especially apparent in what seems to be the profligate wastage of life in nature, Dillard concludes: "Evolution loves death more than it loves you or me .... Are my values then so diametrically opposed to those nature preserves?" (176). The conclusion for her is inescapable: "Either this world, my mother, is a monster, or I myself am a freak" (177). The first half of the choice she gives herself lines Dillard up squarely with Fabre, whose observations of the insect world both fascinate him and horrify him, and they point to the conclusion that nature, and whatever spirit that is animating it, appears malevolent. But the second half of her choice is worse: in her concern for others, even insects, and in her egoistic insistence that even on nature's terms she feels like she must mean something, but alas, doesn't, the suspicion is inescapable—there is no meaning. Her concern for herself is mere ego. Her horror at the horrors of the insect world are the musings of a woman who is really nothing more than an inconsequential freak.

The answer to this quandary, one might think, would be to acknowledge that science's explanation of the intricacies of life, especially through the theories of Wallace and Darwin, in their great sweep through Earth and time, have majesty. The Romantic and Transcendentalist belief in the inspiriting presence of God in nature, and God in insects, was knocked off its pedestal by Darwin and the emerging detailed observations of the natural world. Modernism, clinging desperately to what was receding, responded with anxiety and eventually a terrified glimpse into what appeared to be a void. We can follow this movement quite clearly through literary insects.

I'd like to give Archy the last word on insects and literature, because no literary insect would be quicker to call "humbug" on grand pronouncements about spirit and the majesty of nature than the sarcastic, world-weary, suicidal, delightful cockroach. In "the futility of literature" (Marquis 2006: 307–9), Archy presents a much different take than Frost on the spider/insect encounter. A fly, arguing with a spider over the fly's incipient demise, claims that he performs an important function in the world by spreading disease and ridding the world of "old soaks who | have weakened their systems | with liquor and iniquity." The spider responds that in his web spinning, he is an artist. "i do not | serve utilitarian deities | i serve the gods of beauty | look at the gossamer webs | i weave they float in the sun | like filaments of song." The fly responds: "you have convinced me | said the fly say no more | and shutting all his eyes | he prepared himself for dinner."

Archy reports further:

and yet he said i could
have made out a case
for myself too if i had
had a better line of talk

of course you could said the spider
clutching a sirloin from him
but the end would have been
just the same if neither of
us had spoken at all

boss i am afraid that what
the spider said is true
and it gives me to think
furiously upon the futility
of literature

CHAPTER EIGHT

Insects in Art

A Planet in Peril and the Insect Muse

BARRETT A. KLEIN
AND TIERNEY R. BROSIUS

Insects have a long, rich history of inspiring artists. Their paradoxical nature makes them particularly suited to the task. Insects are constantly underfoot, yet are poorly understood. They are vital to the overall health of our ecosystems, yet as vectors of disease they are the most lethal animal taxon (World Health Organization 2014). They are essential for crop production, yet billions of dollars are spent each year controlling them in our agricultural systems. Insects are often considered subjects of splendor and beauty, yet they are also the source of great fear and disgust (Lockwood 2013). From ornament to symbolic totem, insect imagery spans the entirety of human culture. Here, we examine how the most biodiverse multicellular organisms on the planet are being used by contemporary artists in the modern era. Because we could not hope to give a complete overview of insect-inspired artwork in the modern era, we instead focus on the topic of greatest global concern: the environmental plight of the planet. There is no doubt that the modern era will be defined by the current environmental crisis, evident in a recent, rapid loss of species. How is the loss of biodiversity caused by humans being addressed by artists and, more specifically, how are artists using the most speciose group of animals to reflect this loss? What unique role does something so commonly dismissed as insects serve in modern and contemporary art?

THE ENVIRONMENTAL CRISIS

Scientists have warned for more than thirty years that we are at the start of the sixth mass extinction, and this time the change is anthropogenic. Some call the current geological epoch the Anthropocene in anticipation of the long-term consequences of our environmental destruction. Current estimates put background extinction rates at 114 times that of the pre-human background rate (Ceballos et al. 2015), a veritable "biological annihilation" (Ceballos et al. 2017). The environmental crisis specifically will be felt in terms of the immense loss of biodiversity. Massive reductions in biodiversity will be catastrophic not only in terms of the irreversible loss of "endless forms most beautiful" (Darwin 1859: 490), but also in terms of the impact these reductions are predicted to have on the quality of human lives (Cardinale et al. 2012). Biodiversity is the underpinning of global health (Cardinale et al. 2012), and many factors are contributing to its loss, including climate change. Climate change is predicted to have severe, irreversible global effects that will "extend beyond the entire history of human civilization thus far" (Clark et al. 2016). At this moment, the mechanisms driving climate change are being debated, despite mountains of scientific evidence that point to human activity as a primary underlying cause for the rate of change. Unless we make substantial changes in our behavior, the loss of biodiversity and to humankind is predicted to be detrimental to human and environmental health.

It is clear that warnings given by the scientific community alone are not enough to begin the necessary, massive cultural shift. The messages of doom and gloom have done little to elicit a response in the last fifty years. Scientists are trained to communicate within their discipline, but too few are effective at communicating science to the general public. Even with recent increased focus on science communication and despite the urgency of the message, effective communication of climate change is lagging and lacking. Some have pointed to the effectiveness of art in promoting environmentalism (Hayman 2016).

> Even the most proficient scientist would not succeed in communicating the severity of an environmental problem to the larger public, because she/he would not know how to evoke a mixed response from the audience, or provoke a deep sense of belonging or familiarity and responsibility on a more regional and local level. This can be the role of an artist, a director or a writer. Art has the power to impact the environment, as well as raising awareness of environmental issues beyond charts and statistics.
>
> (King 2010)

It is not surprising that art has been used as a tool in publicizing the environmental crisis. Human culture has a long history of using both art and science to make sense of its surroundings. Humans' relationship with nature was likely the first theme to appear in art, and nature has continued to inspire artists throughout history. The depiction of nature offers the opportunity to examine cultural attitudes about nature and how those attitudes have changed over time. Humans have experienced a transition in the last ten thousand years of being at the complete mercy of nature to slowly gaining an understanding of how to mitigate, if not control, the impact of natural forces. Art has changed accordingly. In the last century, we have seen a drastic change in themes within art inspired by nature. Many artists have shifted from using art to understand the essence of nature to depicting its fragility in light of human activity (King 2010; Brown 2014: 6–15). This shift overlaps with the establishment of an environmental movement.

The environmental movement has its roots in the 1960s, with Rachel Carson's publication *Silent Spring* (Carson 1962) serving as the rallying point for social, scientific, and political change. The historic central themes of humans and nature in art transformed as evidence began to mount that anthropogenic changes could have large-scale negative effects on human and environmental health. While a definition of "environmental art" is still evolving, a central theme is the examination of the impacts humans have on nature. More specifically, we can examine environmental art as work that attempts to raise awareness about the fragility of the environment (Thornes 2008), or art that pursues environmental and conservationist agendas (Pacquement and Williams 2003). Environmental art addresses a diversity of environmental topics using a wide array of methods and media. In this chapter, we investigate how insects are used by artists to convey the loss of Earth's biodiversity.

Because insects are the most biodiverse multicellular organisms, they are particularly well suited to investigate themes within biodiversity loss. Surveying relevant insect art, we found that human involvement in environmental distress fell largely into the following, non-mutually exclusive categories: 1) habitat destruction/change, including climate change; 2) invasive species; 3) pollution, including the use of pesticides; 4) decline of pollinators, including colony collapse disorder (CCD); 5) intentional extermination or modification (e.g., bioengineering) of insects; and 6) concern for the environment/insects. The first three of these categories overlap with categories of threats to biodiversity adopted by Edward O. Wilson (2002), summarized in the acronym HIPPO: **H**abitat loss, **I**nvasive species, **P**ollution, human over-**P**opulation, **O**verharvesting by hunting and fishing. Interestingly, we found almost no work that specifically addresses overpopulation or overharvesting. We will return to this observation in our conclusion. Using our categories of environmental threats as a template, below are examples of artists presenting new ways of examining the environmental crisis by employing insects.

HABITAT DESTRUCTION

> Unsurprisingly, forests play second fiddle to human demands for agriculture and urbanization. Intellectually, we recognize that forests are the lungs of the planet, but not enough is being done to protect this precious resource. Virtually every insect on the endangered species list is there because of loss of habitat. Many people who visit my exhibitions were never aware that such unusual insects exist.
>
> (Angus 2018)

Habitat destruction (or change) is the most visually apparent evidence that humans are adversely affecting the environment. Intensive use of resources and climate change are rapidly altering ecosystems globally. Because habitat destruction results in large-scale changes in the landscape, environmental art frequently focuses on this destruction in visually striking and emotionally profound ways.

Though insects' dimensions are minute and, individually, most insects often go undetected by the human eye, some artists have used insects' small size to make viewers consider human-induced damage on a global scale. For example, Catherine Chalmers filmed hordes of leaf-cutter ants defoliating trees as a metaphor for humans' capacity to destroy habitat in *We Rule* (2013) (Matilsky 2018). Jennifer Angus hopes to increase awareness of environmental issues when assembling thousands of pinned insects in large, site-specific installations (Angus 2018). As with Ilya Kabakov's *The Life of Flies* (1992), Tessa Farmer's *The Perilous Pursuit of a Python* (2013), and Jan Fabre's beetle assemblages, there is something strange about seeing a purposeful arrangement of many insects indoors. The contrast between our normal interactions with insects and placement within an unexpected context (e.g., Angus's geometric arrays on gallery walls) can cause a viewer to pause and change perspective.

Few natural aggregations inspire compassion and awe like those of migrating monarch butterflies. Every winter, millions of monarchs journey to Mexico to escape cold temperatures, but this marvel is threatened by climate change and habitat destruction. Monarch butterflies have experienced a major decline in population size over the last several decades (Semmens et al. 2016; Thogmartin et al. 2017), and habitat loss in both their summer

FIGURES 8.1a, 8.1b *In the Midnight Garden* (2015), installation by Jennifer Angus for the "Wonder" exhibit, Renwick Gallery, Smithsonian American Art Museum. Walls (top) were colored with cochineal extract (Hemiptera: *Dactylopius coccus*) and adorned with pinned insects (detail, bottom). Angus has expressed her desire to increase awareness about habitat destruction. Photographs by Roy Blunt, reproduced with permission from the artist.

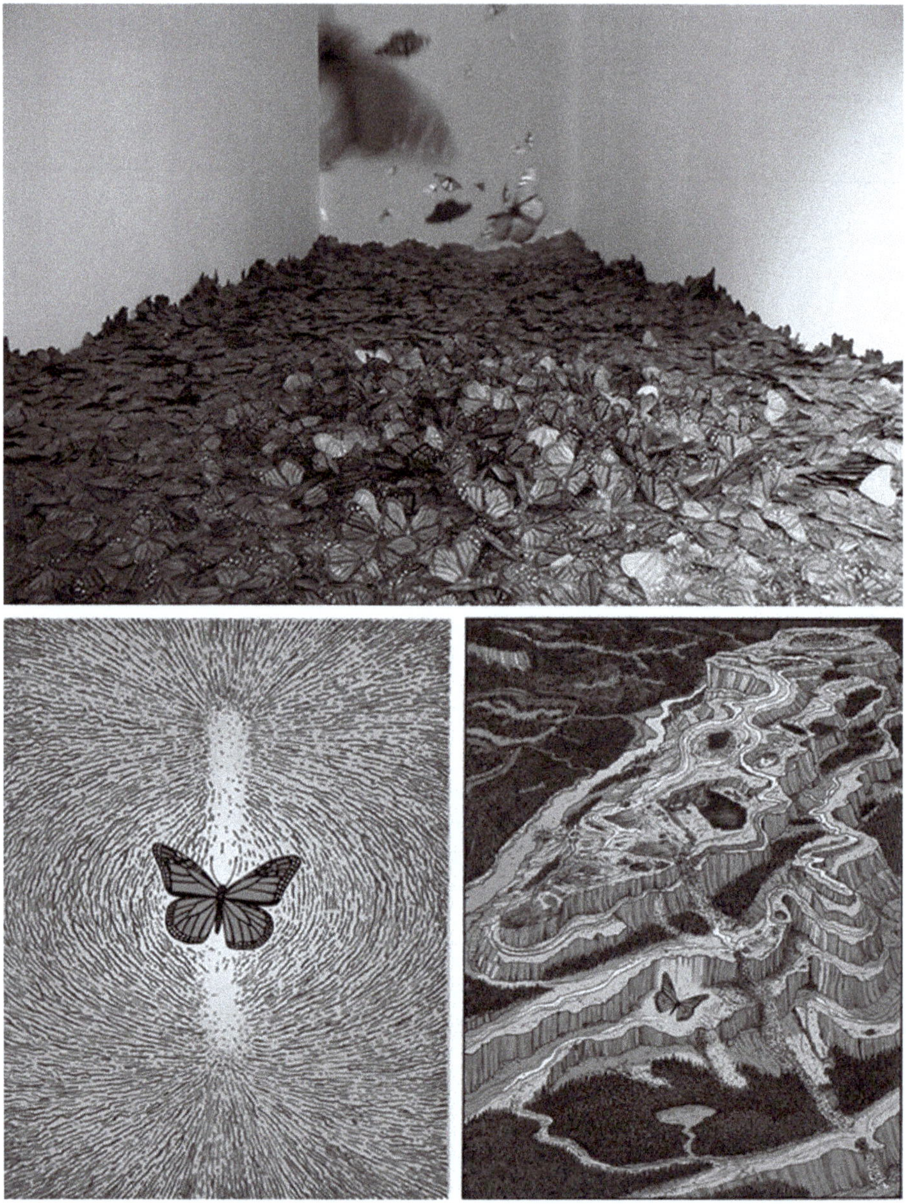

FIGURES 8.2a, 8.2b, 8.2c Art addressing the threat of habitat destruction on monarch butterfly populations. Top: Erika Harrsch's *Eros-Thanatos*, video and sound installation with sixty thousand cut-out butterflies printed in polycarbonate (2011 at *Papillonnages*, Centre d'Art Contemporain du Luxembourg Belge, CACLB, Site de Montauban-Buzenol, Belgium). Bottom: two pages from *Ruins* (2015), a graphic novel by Peter Kuper highlighting the treacherous migration of monarch butterflies (Lepidoptera: *Danaus plexippus*). Reproduced with permission from the artists.

breeding grounds and overwintering areas has been cited as a major cause for their decline (Brower et al. 2011; Flockhart et al. 2014). The struggle of the monarch has captured the attention of artists Erika Harrsch, Asuka Hishiki, and Peter Kuper. Harrsch harnessed the wonder of such perilous migrations in her work *Eros-Thanatos* (2006), in which video and sound accompanied sixty thousand polycarbonate-printed cutouts of monarchs, appearing dead on the floor, to be handled and trampled by visitors. Hishiki installed a piece that included thousands of hand-printed monarch butterflies on recycled paper for another captivating gallery phenomenon—in *Arch of Monarchs* (2013), butterflies appear to flutter from the ceiling. While bringing attention to an ecological concern, both Harrsch and Hishiki were likely motivated by the emotional impact of multiplying a single specimen into a massive aggregation. Kuper, an alternative comic artist and illustrator, also chose to use the monarch butterfly as part of his work, dedicating his graphic novel *Ruins* "to pollinators everywhere." *Ruins* features an entomologist as protagonist, and a monarch butterfly as the focal thread, used to connect elements of the storyline. Kuper (2015) discusses monarchs' predicament, both in the fictional story and in the book's closing facts about social and environmental strife.

Habitat modification in the form of urbanization also spells environmental distress, and possible species loss. Angela Thames produced *Surrey Butterflies* (2007–8), works that focus on how urban landscapes strain nearby wildlife as urban environments inevitably expand. She used common butterfly species to highlight changes over time. Mike MacDonald created butterfly gardens and projections of butterflies (e.g., *Touched by the Tears of a Butterfly*, 2001), and expressed a concern that habitat modification, specifically around Winnipeg, could be displacing butterflies (MacDonald 2004: 218). The Beehive Collective produces environmentally and politically minded narrative graphics in which insects are agents of, for example, *The True Cost of Coal* (2010), warning of habitat destruction and climate change.

Climate change, considered the largest-scale threat to habitats, has become a prevalent theme in art, and at least six posters created by Marina Zurkow, Una Chaudhuri, Oliver Kellhammer, and Fritz Ertl prominently feature insects for the purpose of challenging viewers to think about climate change. *Dear Climate* (2014–ongoing) includes one poster with the slogan "Climate denial is a hoax," and another "Spend quality time with an insect," the latter featuring a cockroach.

FIGURE 8.3 *Dear Climate* (2014–ongoing), by Marina Zurkow, Una Chaudhuri, Oliver Kellhammer, and Fritz Ertl, featuring cockroach and climate change commentary. Reproduced with permission from the artists.

INVASIVE SPECIES

> It's very easy to kind of blame the organism, right?—to make a villain of the plants and animals that are invading other parts of the world, but of course it is us—humans—who get them there. We are really just taking a gamble by not really doing anything. Like all gambles, there will be winners and losers, but probably there will be more losers than winners.
>
> (Mark Dion; youtube.com/watch?v=rohi41depw8)

Invasive species constitute one of the most overlooked causes of environmental damage and biodiversity loss. An analysis of data taken from the IUCN Red List (International Union for the Conservation of Nature) has revealed that invasive species represent the second most common threat that has led to species becoming extinct (Bellard et al. 2016). Many insects certainly suffer from the introduction of invasive species, and Reinhard Reitzenstein alluded to this in *Memory Vessel* (1999), a Chilean installation in which beeswax-lined, bowl-shaped carvings symbolized the threat to bees by toxic, groundwater-tapping, invasive eucalyptus trees. Insects are better known as the harbingers of destruction, however, often filling the insidious role of invasives themselves. Accidental introductions by humans, like of the moth *Lymantria dispar* and the emerald ash borer beetle (*Agrilus planipennis*), have cost communities both in the destruction of millions of trees and the cost of recovering from the loss of large numbers of urban greenscapes. Invasive insects appear cryptically among other invasives in Isabella Kirkland's *Ascendant* (2000), and are prominently featured in Mark Dion's exhibit *Harbingers of the Fifth Season* (2014). In Dion's installation, the viewer sees what looks like the work desk of an artist intensively examining the spread of invasive species, displayed on a bulletin board. On the reverse are images of animals that have gone extinct, in some part due to invasive species.

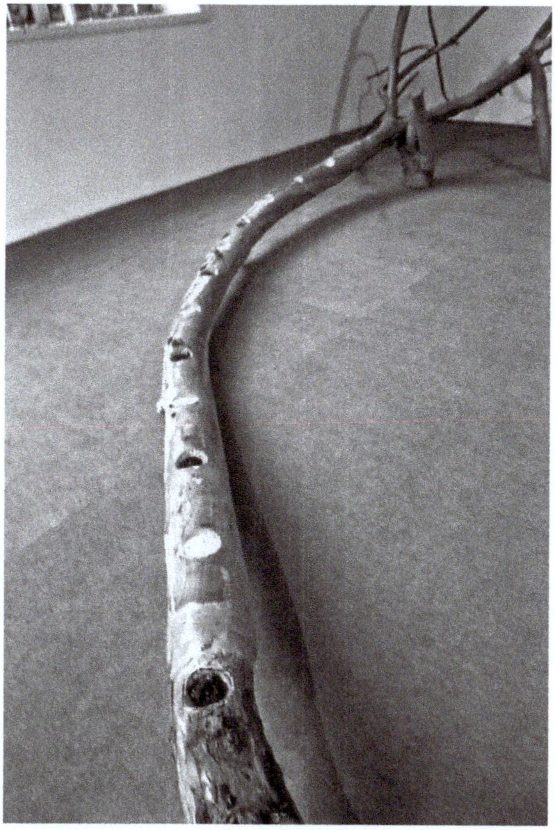

FIGURE 8.4 *Memory Vessel* (1994), an installation of invasive eucalyptus with beeswax-lined bowl-shaped carvings, by Reinhard Reitzenstein. Reproduced with permission from the artist.

POLLUTION

> Our schools and universities produce adults who cannot see. They look but see nothing. I call them see-blinds. The consequence is, we cannot have a good functioning democracy with see-blinds people. They don't see the litter, the pollution, the power structure. And of course, they are absolutely blind to what we do to the world of the insects, to nature and thus to our children. The deformed insects, or true bugs, visualize what we do.
>
> (Cornelia Hesse-Honegger in Aloi and Heese-Honegger 2009)

> As we look to further isolate ourselves from the outside world, we in turn hold the fate of all creatures, great and small, at our disposal. But in the end, these pests become more resistant to the poisons of mankind. Able to mutate these resistances into stronger, more impervious species, these life forms march forward, leaving mankind to live with the poisons they have created.
>
> (David Prochaska; davidprochaska.com)

Even when habitats are otherwise suitable to support organismal diversity, the threat of introduced environmental pollutants causes biodiversity loss. Insects offer a particularly compelling way to inspire artists to call attention to the dangers of pollution, as some species can serve as bioindicators, warning us of environmental change (McGeoch 1998). Insects are among the first organisms to respond to unintended environmental pollution due to their presence at all trophic (feeding) levels above the photosynthesizing producers (e.g., plants). Populations of bioindicators are monitored for their diversity and abundance as a barometer of environmental health. The short generation times of insects allow us to consider the multi-generational and long-term effects of human actions. Cornelia Hesse-Honegger's work uses mutated insects as bioindicators, painting intimate and detailed images of individual insects found in landscapes exposed to pollutants. Her work studies deformed insects, thought to be irradiated mutants of nuclear fallout (e.g., Chernobyl, Fukushima) and "safe" nuclear sites (Hesse-Honegger 1996; Kemp 1998; Hesse-Honegger 2001; Hesse-Honegger and Wallimann 2008; Körblein and Hesse-Honegger 2018), or victims of residual Agent-Orange contamination (Vietnam) (Hesse-Honegger n.d.). Hesse-Honegger's paintings have appeared on silk (Hesse-Honegger 2008), and inspired the designers Lanzavecchia and Wai to depict fictive, irradiated insect mutants in their nuclear MUTAZIONI carpets, *Tacua Fukushimae and Amaurodes Chernobilis* (2013, https://www.designboom.com/design/lanzavecchia-wai-mutazioni-carpets-for-nodus//). Hesse-Honegger's images are subtle and haunting, and this may be due to her analytic, scientific approach to rendering insects that display jarring deviations from normal, natural symmetry. Hesse-Honegger's works show how studying the smallest creatures around us we can see how unintended pollution is altering the natural landscape.

The not-so-subtle use of insects in the 1950s "big bug movies" also served as a warning of the threat of nuclear technology, as well as of the expanding use of synthetically produced chemicals. Giant, irradiated insects embodied the fears of unknown consequences resulting from the overuse of new technology (Leskosky 2007; Tsutsui 2007; Bellin 2009). While these films are thought to have reflected many cultural concerns of the era (such as the Cold War), giant insects seem to reflect a societal awakening to our newly discovered power over nature, while at the same time our fear of becoming powerless to control what we have created. Advances in organic chemistry, genetics, and nuclear energy were both exciting and terrifying in scope. Science was both the cause and often the solution to the giant insect problems featured in these films (Bellin 2009). Art works, in this case science fiction motion pictures, help society reflect on contemporary concerns, and tiny insects magnified into giants are—like Angus's, Chalmers's, Harrsch's, and Hishiki's massive aggregates of insects—reliable subjects for eliciting extreme responses.

Artists wishing to alert the public to the danger of more familiar poisons, like insecticides, often turn to insects as well. This may be surprising, considering that much of the pollution that grabs our attention is the collateral damage caused by chemicals used to kill insects. Widespread use of industrially produced pesticides started after the Second World War (Pedigo and Rice 2014: 369), and DDT quickly became one of the most powerful insecticides the world had ever seen. It did not take long, however, before DDT's failures became evident. There was increasing evidence of environmental damage, human health risks, and reports of insecticides becoming less effective due to insect resistance. Not surprisingly, there was a heightened sense of distrust towards the scientists and industry that produced these chemicals. The organic food market is a response to consumers' wariness of pesticides on our bodies and on the environment. Christy Rupp has been making art to raise environmental awareness since the 1970s, and pollution plays a central role in her works featuring insects, including *Glyphosate (Roundup) Molecule and Cotton Boll Weevil* (1999). David Prochaska's *Point Arrow to Dot* (2003–6) features oil painting portraits of "pests" adorning individual cans of insecticide.

Recent public concern over pesticide use has been in response to their effects on pollinating insects. Many artists expressing concerns over pesticides affecting pollinators have focused on colony collapse disorder (CCD), a mysterious phenomenon associated with honey bees in which a majority of honey bee workers vanish from hives. It is likely that CCD is not caused by a single factor, but the compounded effect of several threats. One of the most publicly visible

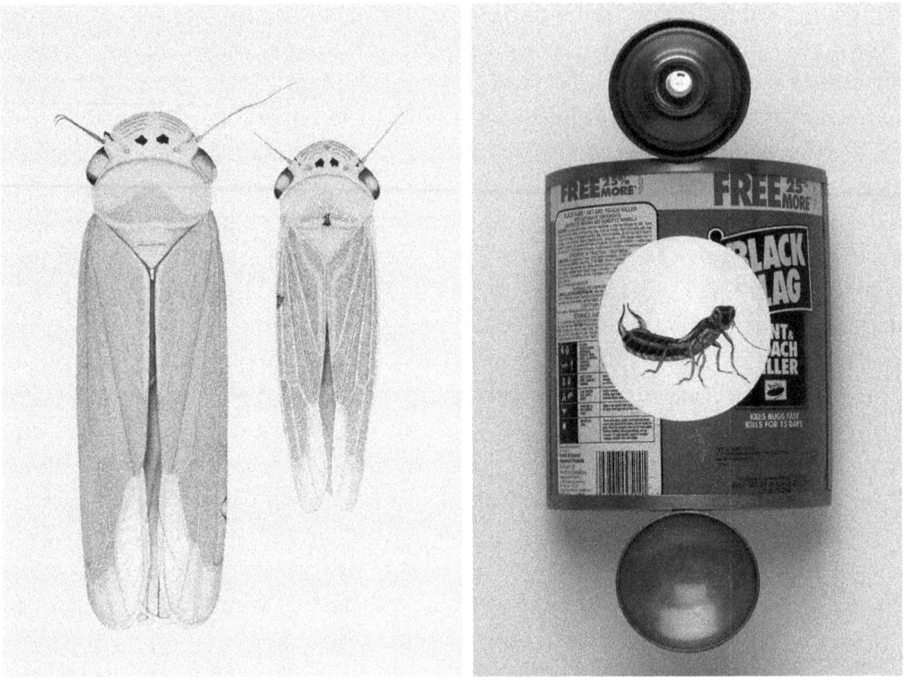

FIGURES 8.5a, 8.5b Pollution. Left: *The asymmetry of two leafhoppers from Shinobu Spa, near Fukushima City, Japan* (2017) is the most recent of Cornelia Hesse-Honegger's evidence of the dangers realized by radiation exposure near nuclear power plants (Hemiptera: *Cicadella viridis*). Specimens were collected after the Fukushima Daiichi nuclear power plant disaster. As described by Hesse-Honegger: "The left leafhopper has two feelers on the left side and a dark patch on the right wing. The right leafhopper is too small with a growth on the thorax." The pale coloring was unique, in Hesse-Honegger's experience, to the area of the Fukushima fallout (Hesse-Honegger, personal communication). Right: *Point Arrow to Dot* (2003–6) is one of a series of insecticide cans adorned with an oil painting portrait of an arthropod (earwig, Dermaptera) by David Prochaska. Reproduced with permission from the artists.

causes has been the use of pesticides, particularly neonicotinoids. Evidence has indicated that neonicotinoid insecticides weaken the immune systems of bees, causing increased vulnerability to parasites and pathogens, such as the varroa mite (Brandt et al. 2016; Sánchez-Bayo et al. 2016). The art of Kindra Crick, Sarah Hatton, and Harriette Tsosie examine the consequences of pesticide use on bees. Crick's *Lost II* (2015) raises concerns about pesticide effects on a bee's learning and memory. Hatton's *Bee Works* series (2013–15) makes use of thousands of dead bees arranged in mathematical patterns to raise concerns over pesticide use. Finally, Tsosie's *The Dead Bee Scrolls Triptych* (2015) is a symbolic hive, collapsing due to the actions of pesticides, depicted as molecules across the surface of encaustic (hot beeswax painting) scrolls.

DECLINE OF POLLINATORS AND CCD

> The bee crisis is worldwide, and is as serious a threat as global warming, and yet it is not nearly enough of a household issue. I hope to spread awareness of bee colony collapse to a broad audience with these works, in a conceptual way that elicits a physical and emotional response in people.
>
> (Sarah Hatton; sarahhattonartist.com)

The decline of pollinators has been a growing concern in the twenty-first century. In the United States alone, pollination by honey bees and other insects in 2010 was valued at $29 billion USD (Calderone 2012). The loss of this vital service could undermine food security for millions of people worldwide. Artists have used insect pollinators to highlight the underestimated value of pollinators and their relevance to our survival. Jenny Kendler's performance art project *Milkweed Dispersal Balloons* (2014–ongoing) seeks to raise awareness of the damage caused by pesticides, habitat loss, and climate change on (largely invisible) insect pollinators. Balloons filled with milkweed seeds are passed out from a traditional food cart, resulting in humans serving as dispersers of plants used as feeding and oviposition sites by monarch butterflies (Aloi 2016).

Most artists concerned about pollinators turn to bees, and exclusively to the western honey bee (*Apis mellifera*), some using actual honey bees or beeswax to create their art (Klein 2013, 2015, 2022). Anne Brodie publicly displayed pieces titled *BEE BOX* (2011), suspending dead bees "flying nowhere, pollinating nothing" (Brodie 2018) to explore the relations between art, science, and society. Bärbel Rothhaar, expressing concern for pollinators, produced *Bee Spaces*, integrating honeycomb with protruding human faces cast in beeswax (Brückner and Rothhaar 2010). Matt Willey chooses to advocate for pollinators by painting murals across the United States (Burnham 2018).

In the 2015 exhibit "The Buzz Stops Here," artists including Kindra Crick and Harriette Tsosie (mentioned above) contributed works of encaustics, a technique using heated, pigmented beeswax, to convey their concern for bees. In *Collapse 5* (2008) and others in this series, Erin Anfinson imaginatively depicted the mysterious culprits of colony collapse disorder by creating "something absurd and elusive," resulting in fantastical aberrations of bees (Gillespie 2015). These pieces reflect the uncertainty and complexity surrounding the disappearance of pollinators. Tsosie (in a personal communication) posits an additional layer of complexity to the colony collapse disorder phenomenon, and a purely pragmatic perspective as to why encaustic artists might care: fewer honey bees means less beeswax, and likely rising costs for the material that defines the encaustic medium.

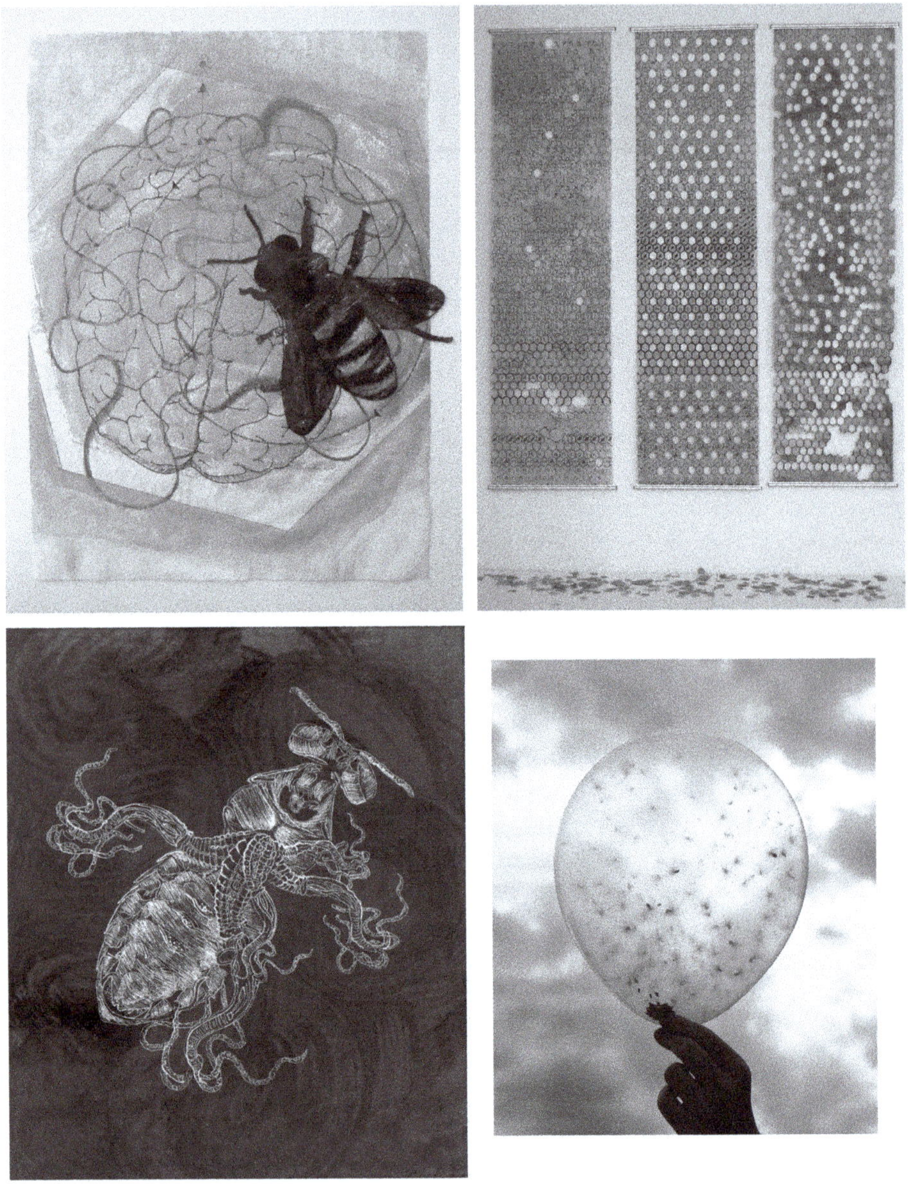

FIGURES 8.6a, 8.6b, 8.6c, 8.6d Decline of pollinators and concern about colony collapse disorder. Top left: *Lost II* (2015), by Kindra Crick. Top right: *The Dead Bee Scrolls Tryptych* (2015), by Harriette Tsosie. Both of these works specifically warn of pesticides' negative effects on honey bees. Bottom left: *Collapse 2* (2008), by Erin Anfinson. Bottom right: *Milkweed Dispersal Balloons* (2014–ongoing), by Jenny Kendler. Reproduced with permission from the artists.

INTENTIONAL EXTERMINATION OR MODIFICATION OF INSECTS

> Humans are incredibly efficient killers, yet remarkably queasy at facing or acknowledging what we do. For us, there is a disjuncture between mass, anonymous, silent deaths, and those that have been individualized. We do not feel the same emotion and responsibility for what we do not witness, whether it is a behind-the-wall pesticide death, or the graver problem of wildlife loss from habitat destruction. But, for the animal facing extermination or extinction, what meaning are our distinctions?
>
> (Chalmers 2004)

Although less than 1 percent of all insect species are characterized as pests, much of our interaction with insects involves their intentional eradication from where we live, play, or grow our crops. Catherine Chalmers challenges viewers to examine their perceptions on how they place value on animals by using live and dead American cockroaches (*Periplaneta americana*) in works that evoke

FIGURE 8.7 *Hanging*, from the *Executions* series (2003) by Catherine Chalmers, incites questions of empathy and intentional extermination. "The Execution series is not about the suffering humans have endured at the hands of humans, but what other species have endured at the hands of humans" (Hunter and Aloi 2007). Reproduced with permission from the artist.

empathy for their presumed suffering. Cockroaches, nearly universally maligned as unwelcome invaders (Gordon 1996), are perfect subjects for Chalmers because she can explore “the aesthetics and emotional parameters that inform whether we embrace nature or hate it” (personal communication). Cockroaches evoke strong emotions (hatred, revulsion) that are out of proportion to the harm cockroaches actually pose, and by staging faux executions, or adorning live cockroaches with iridescent feathers or brightly colored paints, Chalmers esthetically plays with our emotions, and tests our limits of love for nature (Chalmers 2004).

If we are not exterminating organisms, we are modifying them. Artists expressing caution about Frankensteinian departures from nature look to insects to alert us to possible issues associated with genetic modification, or our other means of tinkering with organisms. Richard Pell curates organisms at the Center of PostNatural History, in which the products of science overlap with artists, and become a wondrous mix of both (Pell et al. 2015). The Center is “dedicated to the advancement of knowledge relating to the complex interplay between culture, nature, and biotechnology,” and in it, genetically modified fruit flies (*Drosophila melanogaster*; artist: Soyo Lee), mosquitoes (*Aedes aegypti*), and screwworm flies (*Cochliomyia hominivorax*) provide a context and history of human intervention with nature. For some, genetically modifying insects can be helpful (screwworm fly extirpation) or ethically troubling. The standard practice of genetically engineering mutant strains of fruit flies is highlighted in Alexis Rockman’s *The Farm* (2000) (Natural World Museum 2007).

We pollute Earth’s waters with plastic, altering insects’ freshwater and saltwater habitats. Freshwater caddisfly larvae are now incorporating microplastics in their cases (Ehlers et al. 2019), and ocean-dwelling insects use our sea of plastic refuse for additional oviposition sites (Goldstein et al. 2012). This “*Ecosystem of Excess*” has helped Pinar Yoldas (2015) to imagine “a new Linnean order of post-human life forms” (www.pinaryoldas.info), and create hypothetical changes in her “plastisphere insects” (2014). Fictive modification gets even stranger, as with Christy Rupp’s *Genetically Modified Bugs* series (1999), or Donna Conlon’s *Nature Improvement Project* (2007), in which crane fly specimens (with the help of adhesive) sprout a head full of antennae or a body overwhelmed with legs or wings.

CONCERN FOR THE ENVIRONMENT/INSECTS (HUMAN INVOLVEMENT UNCLEAR)

> A slight change in climate or the disappearance of a single plant species can cause the extinction of one or more species of insects.
>
> (Linsenmaier 1972)

FIGURES 8.8a, 8.8b Modification of insects. Top: *Nature Improvement Project* (2007), one of a series of Frankensteinian crane fly modifications, by Donna Conlon. Bottom: *The Farm* (2000), featuring a genetically modified fruit fly (in circle, between modified chickens), by Alexis Rockman. Reproduced with permission from the artists.

Artists can be more obscure about their intentions or motivations than in many of the examples cited above. Likewise, those who write about or critique art can be equally obscure. Artists can express concern for the environment, or for insects, without explicitly blaming humans, or specifically indicating how humans are a threat. While human involvement may be unclear, it is worth mentioning art with insects that is intended to bring attention to environmental concerns. Our final category can be subdivided into concern for insects, and the disconnect between humans and nature.

Considering insects' roles as pollinators, decomposers, bioindicators, and as key participants in food webs, concern for the wellbeing of insects should come as naturally as the concern about declines of tigers, elephants, or polar bears. It takes a special type of person, however, to recognize and advertise the decline of an insect population. Troubled about the future of bees, Beate Kratt depicted the "essence of the last bee's dance captured in a jar" in *Traces* (2015) (Gillespie 2015). Worried about human consumption and lifestyle choices, Anthony Heywood drew a bumble bee (*Bombus* sp.) and other endangered insects in his *Species* series (Smithers 2008). Even Andy Warhol, generally not recognized for his interest in the natural world, was motivated by activists Ronald and Frayda

FIGURE 8.9 Concern for environment/insects. *Roundup: An Entomological Endeavor for the Smart Museum of Art* (2000), by Mark Dion for the Smart Museum of Art. Dion challenges us with insects to realize that we are never truly beyond nature, even in a museum gallery. Reproduced with permission from the artist. Photograph © 2017 courtesy of The David and Alfred Smart Museum of Art, The University of Chicago.

Feldman to produce a pop-art *Endangered Species* series (1983), which includes the Callipe silverspot butterfly (*Speyeria callippe callippe*).

Why care about insects, or nature for that matter, if your connection to both is minimal? Despite evidence suggesting that humans are happier when more closely connected to nature (Nisbet et al. 2011), there is a growing disconnect as dependence on technology grows. Several artists, including Mike MacDonald (mentioned above), have expressed concern about this human-nature disconnect by using insects as subjects in their work. Liao Wenfeng videotaped a disoriented forest ant wandering in a science lab in *From Here to There* (2010). According to exhibit curator Stefanie Dathe, "In light of the increasing and predominant technology and globalization of the human race, the ant's useful and indispensable existence is ignored" (Hoenes-Stiftung and Dathe 2012: 90–1). Jaune Quick-to-See Smith's art embraces Native American philosophy, and in *Carousel* (2004), fears of living unsustainably are represented in a piece vividly portraying insects. In *Roundup: An Entomological Endeavor* (Smith 2001), Mark Dion claims, "Even in the neutral white cubes of our art

FIGURE 8.10 Concern for environment/insects. *Untitled* (1985), the comb-laden apisculptural aftermath of massive bee activity, orchestrated by Garnett Puett. Writer Timothy Cohrs, responding to the reaction of an audience to this public piece, reported a worrisome disconnect between humans and nature. Reproduced with permission from the artist.

museums nature, in the form of insect life, challenges our view that human culture is somehow outside or beyond nature."

Finally, Garnett Puett has collaborated with honey bees to produce "apisculptures," in which his waxen forms are reworked by comb-building worker bees. After one of Puett's human figures was altered and became a human-nest hybrid (*Untitled*, 1985), writer Timothy Cohrs (1985) reported:

> The obvious metaphor here is that mankind is so out of whack with nature, and its perception of nature so screwy, that simple natural products and processes register as horrible deformations ... Puett's comb-laden figures evoke the metaphor of biological deformity resulting from a fatal lack of environmental integration.

CONCLUSION

Choices we make affect biodiversity, and a prevalent theme in modern era art asks us to reflect on those choices. Our urges for material wealth (reinforced by a consumer culture), paired with our psychological inability to grasp the enormity and uncertainty of climate change, may lead to the end of a large portion of our planet's biodiversity (Gifford 2011; van der Linden et al. 2015). No environmental action based on science alone can change this trajectory. Conservation must have a philosophical and ethical foundation that moves beyond viewing nature as being separate from, or created for, humans (Samways 2005). There are philosophical movements that view humans as part of a global ecosystem, and nature as more than purely utilitarian, but those viewpoints have yet to become mainstream in the largest economies of the world (Naess 1980). The necessary shift in cultural values that will need to take place may depend on the arts.

> Artists can help awaken us. They can recall to us our place on this planet, remind us that we breathe the same air the hummingbird breathes, the same air that contains oxygen released by our trees and the same air the dolphin inhales before traveling to the deep seas. The sacred aspect of our relationship to nature has to be accessed, enriched and expressed, enlivening our senses and intellect. Artists help us change the way we see.
>
> Mia Hanak in Natural World Museum 2007:170–1

Our survey and categorization of environmentally minded insect art came with several revelations. Fewer examples of invasive species as threats to biodiversity exist in art than this category should inspire, and we found almost no insect art works that specifically address human overpopulation or overharvesting as threats to biodiversity. The mismatch between invasive species-themed insect art and the threat that invasive species pose may indicate the unsettling prospect that humans (and artists, specifically) are generally less aware of, or less impressed by, the threats posed by invasive species. If so, there exists an opportunity to improve communication between scientist and artist, directly or indirectly. Works addressing environmental issues of human overpopulation exist (Matilsky 2018), so it is more difficult to speculate about the near absence, in our sample, of insect art involving overpopulation and overharvesting as threats to biodiversity. The ubiquity of insects, the abundance and density of insects, the rapid proliferation of insects, all suggest that insects could metaphorically represent the burden of overpopulation to a receptive artist. Is it that the ubiquitous insect gives the impression that life is inexhaustible, and not sufficiently threatened by our actions to warrant artistic representation? Is overpopulation (and population control) potentially too sensitive a subject due to its globally and culturally complex nature? Or is it that we are reluctant to relate to insects—those alien entities all around us?

It is possible, of course, that our speculations suffer from sampling error/ inadequacy, and there may exist many examples of insect-themed art addressing dangers of invasives, overpopulation, or overharvesting. Attempting to locate and decipher all relevant art is as futile as capturing and identifying every species of insect. In an attempt to expose how underrepresented our sample may be, we asked artists whose insect-based art is not obviously relevant to our project to talk about their art in the context of environmental distress, if appropriate. Although explanation of our focus may have biased artists' responses, worthwhile discussions followed. Contact with several artists resulted in previously unpublished or undisclosed interpretations of their art that were perfectly relevant to our focus. Without inquiring about her motivations, Donna Conlon's *Nature Improvement Project* series appears to fall within a subcategory of insect art less relevant to our theme—that of artificially constructed insect chimaeras. Reaching out to Conlon, however, revealed her motivation in this work to address how misguided humans' manipulation of nature can be. Likewise, before communicating with Mark Dion about his installation *The Great Munich Bug Hunt* (1993), we were unaware that his exhibit was an exposé of German forestry's "open war on insects," due to Germany's policy of treating forests like tree farms. Carim Nahaboo, a scientific illustrator who focuses on insects, informed us of his inclination to draw insects rendered endangered or extinct by human activity. We had the luxury of communicating directly with several contemporary artists, but had to rely on written records for others. What fears, what apprehensions, what environmentally minded inspirations moved many artists to produce their insect art will never be known. Among the decipherable and deciphered others, we will continue to uncover potent examples of art in which insects alert us to the folly of our environmental policies (see Klein and Brosius 2022 for survey results).

Unlike science, art's purpose is not to supply answers. One goal is often to slow down the viewers enough to ask their own questions and reconsider preconceived notions about their surroundings. The artists featured here have moved past the obvious use of charismatic megafauna and the sublime splendor of immense landscapes. Insects serve unique roles within environmental art, and artists draw on their ubiquitous yet mysterious nature. Insects are viewed as compellingly paradoxical; some can serve as useful bioindicators, and others exhibit the perils of invasiveness. Insects are also unique in that, although small, they can be disarmingly gigantic in aggregate, or evoke extreme reactions if artistically rendered to be individually monstrous. Insects provide a perfect opportunity to surprise the viewer with their intrinsic value and beauty, given that a psychological aversion to insects is so common. By examining environmentally focused artwork featuring insects, viewers are challenged to consider how some of the smallest multicellular organisms can be of vital importance to our ecology and to our future survival.

NOTES

Chapter 2

1. The accounts of entomological warfare in this chapter are derived from Lockwood (2009). In particular, the descriptions of insects being used as weapons in the US Civil War comes from Chapter 6, the First World War from Chapter 7, the Second World War Japanese program from Chapters 8–11, the Second World War German program from Chapter 12, the Second World War Allies programs from Chapter 13, the Korean War events from Chapters 14–16, the Cold War US program from Chapter 17, the Vietnam War events from Chapter 18, the Cuban events from Chapters 19–20, the contemporary acts of terrorism from Chapter 23, and the future uses from Chapters 25–6.

Chapter 6

1. The accounts in this chapter of insect tricksters and Charley Patton, Blind Lemon Jefferson, and Furry Lewis are derived from Peterson (2007), Peterson (2013), and Peterson (2014), respectively.

BIBLIOGRAPHY

Abbott, L. and Seroff, D. 2017. *The Original Blues: The Emergence of the Blues in African American Vaudeville*. Jackson, MS: University Press of Mississippi.

Abrams, M.H. 1971. *Natural Supernaturalism: Tradition and Revolution in Romantic Literature*. New York: W.W. Norton.

Aichele, G. and Walsh, R. 2011. "Metamorphosis, Transfiguration, and the Body." *Biblical Interpretation* 19: 253–75.

Aloi, G. 2016. "Milkweed Dispersal Balloons." *Antennae: The Journal of Nature in Visual Culture* 35: 57–69.

Aloi, G. and Hesse-Honegger, C. 2009. "Cornelia Hesse-Honegger: Heteroptera." *Antennae: The Journal of Nature in Visual Culture* 11: 29–36.

Alperovitz, G. 1969. "How Did the Cold War Begin?" In B.J. Bernstein and A.J. Matusow (eds.), *Twentieth Century America: Recent Interpretations*, 390–398. New York: Harcourt, Brace & World.

Anelli, C.M., Krupke, C.H., and Prasad, R.P. 2006. "Professional Entomology and the 44 Noisy Years Since Silent Spring, Part 1." *American Entomologist* 53(1): 224–33.

Angus, J. 2018. "Jennifer Angus." Available online: http://www.jenniferangus.com/ (accessed March 16, 2018).

Annand, P.N. 1947. "Preventive Entomology." *Journal of Economic Entomology* 40(4): 461–8.

Anonymous. 1921a. "Scientists' Battle-line Against New Insect Enemies." *New York Times* (July 31), p. 73.

Anonymous. 1921b. "Says Insects are Menace to Man." *New York Times* (December 28), p. 6.

Anonymous. 1922. "Control of Yellow Fever." *New York Times* (March 5), p. 84.

Anonymous. 1923a. "War in Mosquitos Cuts Down on Malaria." *New York Times* (February 16), p. 30.

Anonymous. 1923b. "Topics of the Times. Hunting Bugs from an Airplane." *New York Times* (June 20), p. 18.

Anonymous. 1926a. "Humans Face Insect War." *New York Times* (June 20), p. 17.

Anonymous. 1926b. "Brief Reviews. The Insect World. How Insects Live: An Elementary Entomology by Walter Housely Wellhouse." *New York Times* (September 12), p. BR18.
Anonymous. 1926c. "International Insect Allies." *New York Times* (September 21), p. 28.
Anonymous. 1928. "Insects Dominate Man, Says Entomologist." *New York Times* (August 14), p. 24.
Anonymous. 1929. "Malaria Curbed in the Canal Zone." *New York Times* (July 5), p. 23.
Anonymous. 1931a. "Armies of 15,000,000 Parasites Mobilizing for Offensive on Connecticut Peach Moth." *New York Times* (March 1), p. N2.
Anonymous. 1931b. "Calls Man to War on Insect Menace." *New York Times* (September 10), p. 22.
Anonymous. 1935a. "A Little Insect Acquires Merit." *New York Times* (March 17), p. SM21.
Anonymous. 1935b. "Government Insect Curbs Prickly Pear." *New York Times* (May 5), p. E12.
Anonymous. 1938. "Insect Pests Now Travel by Plane." *New York Times* (January 16), p. 66.
Anonymous. 1941. "Disease Germs Destroy Beetles." *New York Times* (October 12), p. D8.
Anonymous. 1945a. "Notes on Science." *New York Times* (July 22), p. 71.
Anonymous. 1945b. "More Light on DDT." *New York Times* (September 16), p. X9.
Anonymous. 1945c. "DDT Spray Called Injurious to Birds." *New York Times* (October 23), p. 10.
Anonymous. 1946. "Warns About Insects." *New York Times* (February 9), p. 27.
Anonymous. 1947a. Advertisement. *New York Times* (May 11), p. X30.
Anonymous. 1947b. "Farmers Warned on DDT." *New York Times* (May 25), p. 52.
Anonymous. 1948. "DDT Wins Nobel Prize." *New York Times* (October 30), p. 14.
Anonymous. 1952. "Mosquitos Developing an Armor Against DDT After 9-Year War." *New York Times* (March 14), p. 25.
Anonymous. 1953a. "New Insecticide Ready." *New York Times* (April 5), p. F7.
Anonymous. 1953b. "Hitting Below the Belt, Scientists Suggest Intestinal Attack on the Common Housefly." *New York Times* (December 9), p. 16.
Anonymous. 1953c. "Nerve 'Poisons' Kill Pests Inside Plants." *New York Times* (December 10), p. 63.
Anonymous. 1962. "Books – Authors." *New York Times* (July 3), p. 21.
Anonymous. 1964. "2,000 Experts Meet to Discuss Insects." *New York Times* (July 9), p. 67.
Anonymous. 1990. "A Different Portrait of Animals: Size Based on Diversity." *New York Times* (December 18), p. C4.
Arnold, A.J. 1996. *Monsters, Tricksters, and Sacred Cows: Animal Tales and American Identities*. Charlottesville, VA: University of Virginia Press.
Arnold, M. 1994. *Dover Beach and Other Poems*. n.p: Dover Publications.
Associated Press. 1962. "Atomic Energy Being Used to Eradicate Screwworm." *New York Times* (October 11), p. 24.
Bahti, M. 1999. *Spirit Animal in the Stone: A Handbook of Southwest Indian Carvings and Beliefs*. Tucson, AZ: Rio Nuevo Publishers.

Batuman, E. 2010. "Kafka's Last Trial." *The New York Times Magazine* (September 22). Available online: https://www.nytimes.com/2010/09/26/magazine/26kafka-t.html (accessed May 10, 2023).

Bel Geddes, N. 1940. *Magic Motorways*. New York: Random House.

Bellard, C., Cassey, P., and Blackburn, T.M. 2016. "Alien Species as Driver of Recent Extinctions." *Biology Letters* 12(2).

Bellin, J.D. 2009. "Us or Them!: Silent Spring and the 'Big Bug' Films of the 1950s." *Extrapolation* 50(1): 145–68.

Benecke, M. 2008. "A Brief Survey of the History of Forensic Entomology." *Acta Biologica Benrodis* 14: 15–38.

Benjamin, A. and McCallum, B. 2009. *A World without Bees*. New York: Pegasus Books.

Bennett, H.Z. 1993. *Zuni Fetishes: Using Native American Objects for Meditation, Reflection, and Insight*. New York: Harper Collins Publishers.

Bernal, I. and Groth, I. 1968. *Ancient Mexico in Color*. New York: McGraw-Hill Book Company.

Berry, W. [1977] 2015. *The Unsettling of America: Culture and Agriculture*. Berkeley, CA: Counterpoint Press.

Beutelspracher, C. 1988. *Las Mariposas entre los Antiguos Mexicanos*. Mexico D.F.: Fondo de Cultura Economica Tezontle.

Bloudoff-Indelicato, M. 2013. "Flesh-Eating Beetles Explained." *National Geographic Voices: Ideas and Insights from Explorers*. Available online: https://www.nationalgeographic.com/animals/article/flesh-eating-beetles (accessed July 16, 2017).

Boppré, M. and Vane-Wright, R.I. 2014. "The Butterfly House Industry: Conservation Risks and Education Opportunities." *American Butterflies* 22(1): 24–43.

Bradley, L. 2016. "How *Bee Movie* Won 2016." *Vanity Fair* (December 30).

Brandt, A., Gorenflo, A., Siede, R., Meixner, M., and Büchler, R. 2016. "The Neonicotinoids Thiacloprid, Imidacloprid, and Clothianidin Affect the Immunocompetence of Honey Bees (*Apis mellifera* L.)." *Journal of Insect Physiology* 86: 40–7.

Brodie, A. 2018. Bee Box 2011. Available online: http://www.annebrodie.com/#/bee-box/ (accessed March 15, 2018).

Brody, J.E. 1967. "Scientists Press War on Insect Pests' Birth Rate." *New York Times* (November 26), p. 81.

Brody, J.E. 1976. "Farmers Turn to Pest Control in Place of Eradication." *New York Times* (August 1), p. 1.

Brower, L.P. 1977. "Monarch Migration." *Natural History* 86(6): 40–53, cover.

Brower, L. 1985. "New Perspectives on the Migration Biology of the Monarch Butterfly, *Danaus plexippus* L." In M. Rankin (ed.), *Migration: Mechanisms and Adaptive Significance*, 29. Austin: University of Texas, Contributions in Marine Science, Supplement 27.

Brower, L.P. 2001. "Canary in the Cornfield: The Monarch and the Bt Corn Controversy." *Orion* 20(2): 32–41.

Brower, L.P. 2015. "The Remarkable Monarch." *Natural History* 123(6): 27–9.

Brower, L.P. and Malcolm, S.B. 1989. "Endangered Phenomenon." *Wings* 14(2): 3–9.

Brower, L.P., Taylor, O.R., Williams, E.H., Slayback, D.A., Zubieta, R.R., and Ramírez, M.I. 2011. "Decline of Monarch Butterflies Overwintering in Mexico: Is the Migratory Phenomenon at Risk?" *Insect Conservation and Diversity* 5(2): 95–100.

Brown, A. 2014. *Art and Ecology Now*. New York. Thames & Hudson Inc.
Bruchac, J. 1992. *Native American Animal Stories*. Golden, CO: Fulcrum Publishing.
Brückner, D. and Rothhaar, B. 2010. *Honigbienen in Kunst und Wissenschaft* (Honeybees in Art and Science). Bremen, Germany: Haus der Wissenschaft.
Buffalo Horn Man, G. and Firedancer, S. 1992. *Animal Energies*. Lexington, KY: Dancing Otter Publishing.
Buhs, J.B. 2004. *Fire Ants Wars: Nature, Science, and Public Policy in Twentieth-Century America*. Chicago, IL: University of Chicago Press.
Burnham, T. 2018. "Downtown the Good of the Hive, the Good of the Bees, and the Good of Us All." *Bee Culture: The Magazine of American Beekeeping*. Available online: https://www.beeculture.com/downtown-the-good-of-the-hive-the-good-of-the-bees-and-the-good-of-us-all/ (accessed on July 31, 2020).
Butler, C.A. 2015. "The Monarchs and I." *Natural History* 123(6): 22–5.
Byrd, J.H. 1998. "What is Forensic Entomology?" *Forensic Entomology*. Available online: https://www.forensic-entomology.com/definition/(accessed July 16, 2017).
Calderone, N.W. 2012. "Insect Pollinated Crops, Insect Pollinators and US Agriculture: Trend Analysis of Aggregate Data for the Period 1992–2009." *PLoS ONE* 7(5): e37235. DOI: 10.1371/journal.pone.0037235.
Camus, Albert. 1948. *The Plague*. New York: Penguin.
Cantor, N.F. 1988. *Twentieth-Century Culture Modernism to Deconstruction*. New York: Peter Lang.
Cardinale, B.J., Duffy, E., Gonzalez, A., Hooper, D.U., Perrings, C., Venail, P., Narwani, A., Mace, G.M., Tilman, D., Wardle, D.A., Kinzig, A.P., Daily, G.C., Loreau, M., Grace, J.B., Larigauderie, A., Srivastava, D., and Naeem, S. 2012. "Biodiversity Loss and its Impact on Humanity." *Nature* 486(7401): 59–67.
Carson, R. 1962. *Silent Spring*. New York: Houghton Mifflin.
Carson, R. 2002. *Silent Spring*. Boston: Houghton Mifflin.
Carter, E. 1988. "Frost's 'Design.'" *The Explicator* 47(1): 2–26.
Ceballos, G., Ehrlich, P.R., and Dirzo, R. 2017. "Biological Annihilation Via the Ongoing Sixth Mass Extinction Signaled by Vertebrate Population Losses and Declines." *Proceedings of the National Academy of Sciences* 114(30).
Ceballos, G., Ehrlich, P.R., Barnosky, A.D., García, A., Pringle, R.M., and Palmer, T.M. 2015. "Accelerated Modern Human-Induced Species Losses: Entering the Sixth Mass Extinction." *Science Advances* 1(5).
Chalmers, C. 2004. *American Cockroach*. New York: Aperture Foundation, Inc.
Chapman, R.F. 2000. "Entomology in the Twentieth Century." *Annual Review of Entomology* 45: 261–85.
Childs, P. 2017. *Modernism*, 3rd edn. New York: Routledge.
City of Pacific Grove. 2017. Available online: www.cityofpacificgrove.org/visiting/monarch-butterfly-sanctuary (accessed November 7, 2017).
Clark, F. and Illman, D.L. 2006. "A Longitudinal Study of the *New York Times* Science Times Section." *Science Communication* 27(4): 496–513.
Clark, P.U., Shakun, J.D., Marcott, S.A., Mix, A.C., Eby, M., Kulp, S., Levermann, A., Milne, G.A., Pfister, P.L., Santer, B.D., Schrag, D.P., Solomon, S., Stocker, T.F., Strauss, B.H., Weaver, A.J., Winklemann, R., Archer, D., Bard, E., Goldner, A., Lambeck, K., Pierrehumbert, R.T., and Planter, G. 2016. "Consequences of Twenty-First-Century Policy for Multi-Millennial Climate and Sea-Level Change." *Nature Climate Change* 6: 360–9.
Clarke, S. 2014. "It's Some of Your Beeswax: Bees on Record Labels." *American Entomologist* 60: 23–8.

Clench, H.K. 1964. "A New Species of Riodinidae from Mexico." *Journal of Research on the Lepidoptera* 3(2): 73–80.

Coelho, J. 2000. "Insects in Rock & Roll Music." *American Entomologist* 46: 186–200.

Cohrs, T. 1985. "Garnett Puett's Apisculpture: Uniting the Bee to the Task of Art." *Arts Magazine* 60: 122–3.

Cordry, D. and Cordry, D. 1968. *Mexican Indian Costumes*. Austin, TX: University of Texas Press, The Texas Pan American Series.

Cotter, J. 2002. *Troubled Harvest: Agronomy and Revolution in Mexico, 1880–2002*. Westport, CT: Praeger.

Cullather, N. 2010. *The Hungry World: America's Cold War Battle Against Poverty in Asia*. Cambridge, MA: Harvard University Press.

Daniel, P. 2005. *Toxic Drift: Pesticides and Health in the Post-World War II South*. Baton Rouge: Louisiana State University Press.

Darwin, C. 1859. *On the Origin of Species by Means of Natural Selection, or the Preservation of Favored Races in the Struggle for Life*. London: John Murray.

Davis, F.R. 2014. *Banned: A History of Pesticides and the Science of Toxicology*. New Haven, CT: Yale University Press.

DeBach, P., Huffaker, C.B., and MacPhee, A.W. 1976. "Evaluation of the Impact of Natural Enemies." In C.B. Huffaker and P.S. Messenger (eds.), *Theory and Practice of Biological Control*, 270. New York: Academic Press.

Dillard, A. 1974. *Pilgrim at Tinker Creek*. New York: Harper's Magazine Press.

Dirzo, R. and Garcia, M.C. 1992. "Rates of Deforestation in Los Tuxtlas, a Neotropical Area in Southeast Mexico." *Conservation Biology* 6(1): 84–90.

Dunlap, T., ed. 2008. *DDT, Silent Spring, and the Rise of Environmentalism*. Seattle: University of Washington Press.

Durand, L. and Lazos, E. 2008. "The Local Perception of Tropical Deforestation and its Relationship to Conservation Policies in Los Tuxtla Biosphere Reserve, Mexico." *Human Ecology* 36(3): 383–99.

Dutton, D. 2008. *The Art Instinct: Beauty, Pleasure, & Human Evolution*. New York: Bloomsbury Press.

Eckholm, E. 1986. "Species are Lost Before They're Found." *New York Times* (September 16), p. C1.

Economic Times. 2017. "What is Product?" Available online: http://economictimes.indiatimes.com/definition/product (accessed July 16, 2017).

Ehlers, S.M., Manz, W., and Koop, J.H.E. 2019. "Microplastics of Different Characteristics are Incorporated Into the Larval Cases of the Freshwater Caddisfly *Lepidostoma basale*." *Aquatic Biology* 28: 67–77. DOI: 10.3354/ab00711.

Elson, B. and Gutiérrez G. D. 1999.*Diccionario Popoluca de la Sierra Veracruz. Serie de Vocabularios y Diccionarios Indigenas "Mariano Silva y Aceves*." Número 41. Mexico, DF: Instituto Lingüístico de Verano.

Emerson, R.W. 1996. *Emerson: Essays & Poems*. New York: Library of America.

Enciso, J. 1953. *Design Motifs of Ancient Mexico*. New York: Dover Publications, Inc.

Enciso, J. 1971. *Designs from Pre-Columbian Mexico*. New York: Dover Publications, Inc.

Erdoes, R. and Ortiz, A. 1984. *American Indian Myths and Legends*. New York: Pantheon Books.

Erwin, T.L. 1982. "Tropical Forests: Their Richness in Coleoptera and Other Arthropod Species." *The Coleopterists Bulletin* 36(1): 74–5.

Evans, W.A., Krippendorf, M., Yoon, J.H., Posluszny, P., and Thomas, S. 1990. "Science in the Prestige and National Tabloid Presses." *Social Science Quarterly* 71(1):105–17.

Fabre, J.H. 1991. *The Insect World of J. Henri Fabre*. Boston, MA: Beacon Press.

Fagan, D. and Seltzer, R. 2010. "Frost's DESIGN." *The Explicator* 68(1): 48–50.

Faust, J.L. 1987. "Controlling Insects with Nature's Help." *New York Times* (June 18), p. C3.

Feder, B.J. 1996a. "Out of the Lab, a Revolution on the Farm." *New York Times* (March 3), p. F1.

Feder, B.J. 1996b. "Geneticists Arm Corn Against Corn Borer but Pest May Still Win." *New York Times* (July 23), p. C1.

Feltwell, J. 1986. *The Natural History of Butterflies*. New York: Facts on File Publications.

Ferguson, T.J. and Hart, E.R. 1985. *A Zuni Atlas*. Norman, OK: University of Oklahoma Press.

Field, F. 1974. *Pre-Hispanic Mexican Stamp Designs*. New York: Dover Publications, Inc.

Fisher, A.H. 1929. "Exploring the Hidden Insect World." *New York Times* (June 23), p. SM7.

Fitzgerald, T.D. 2001. "Nightlife of Social Caterpillars." *Natural History* 110(1): 38–43.

Flockhart, D.T.T., Pichancourt, J., Norris, D.R., and Martin, T.G. 2014. "Unravelling the Annual Cycle in a Migratory Animal: Breeding-Season Habitat Loss Drives Population Declines of Monarch Butterflies." *Journal of Animal Ecology* 84(1): 155–65.

Foley, J., DeFries, A.R., Asner, GG.P., Barford, C., Bonan, G., Carpenter, S.R., Chapin, F.S., Coe, M.T., Daily, G.C., Gibbs, H.K., Helkowski, J.H., Holloway, T., Howard, E.A., Kucharik, C.H., Monfreda, C., Patz, J.A., Prentice, I.C., Ramankutty, N., and Snyder, P.K. 2005. "Global Consequences of Land Use." *Science* 309(5734): 570–4.

French Symbolist Poetry. 1958. Trans. C.F. MacIntyre. Berkeley, CA: University of California Press.

Furedi, F. 1998. *The Silent War: Imperialism and the Changing Perception of Race*. New Brunswick, NJ: Rutgers University Press.

Galton, L. 1963. "Great Debate Over Pests – and Pesticides." *New York Times* (April 14), p. 220.

Garrett, L. 1995. *The Coming Plague: Newly Emerging Diseases in a World Out of Balance*. New York: Penguin.

Garrett, L. 2001. *Betrayal of Trust: The Collapse of Global Public Health*. New York: Hachette Books.

Gassmann, A.J., Petzold-Maxwell, J.L., Keweshan, R.S., and Dunbar, M.W. 2011. "Field-Evolved Resistance to Bt Maize by Western Corn Rootworm." *PLOS ONE* 6(7): e22629.

Gauding, M. 2006. *Personal Power Animals for Guidance, Protection and Healing*. London: Octopus Publishing Group, A Godsfield Book.

General Motors. To New Horizons. 1940. Department of Public Relations, General Motors.

Giesen, J.C. 2011. *Boll Weevil Blues: Cotton, Myth, and Power in the American South*. Chicago, IL: University of Chicago Press.

Gifford, R. 2011. "The Dragons of Inaction: Psychological Barriers that Limit Climate Change Mitigation and Adaptation." *American Psychologist* 66(4): 290–302. DOI: 10.1037/a0023566.

Gillespie, H. 2015. "The Buzz Stops Here: An Exhibition of Encaustic Artwork About the Science and Conservation of Bees." Art.Science.Gallery., Austin, TX, USA. Available online: https://issuu.com/artsciencegallery/docs/buzzcatalogue_20150411_1608 (accessed May 10, 2023).

Gilpin, R. 2000. *The Challenge of Global Capitalism: The World Economy in the 21st Century*. Princeton: Princeton University Press.

Glassberg, J., Opler, P., Pyle, R.M., Robbins, R., and Tuttle, J. 1998. "There's No Need to Release Butterflies—They're Already Free." *American Butterflies* 6(1): 2.

Goldstein, M.C., Rosenberg, M., and Chenge, L. 2012. "Increased Oceanic Microplastic Debris Enhances Oviposition in an Endemic Pelagic Insects." *Biology Letters* 8(5): 817–20.

Google Books Ngram. 2018. Available online: https://books.google.com/ngrams (accessed May 10, 2023).

Gordon, D.G. 1996. *The Compleat Cockroach: A Comprehensive Guide to the Most Despised (and Least Understood) Creature on Earth*. Berkeley, CA: Ten Speed Press.

Goulson, D., Nicholls, E., Botias, C., Rotheray, E.L. 2016. "Bee Declines Driven by Stress from Parasites, Pesticides, and Lack of Flowers." *Science* 347(6229).

Grove, D. 2014. *Discovering the Olmecs: An Unconventional History*. Austin, TX: University of Texas Press.

Guiltinan, J. 2009. "Creative Destruction and Destructive Creations: Environmental Ethics and Planned Obsolescence." *Journal of Business Ethics* 89 (Supplement 1): 19–28.

Gullan, P.J., and Cranston, P.S. 2014. *The Insects: An Outline of Entomology*, 5th edn. Oxford: Wiley Blackwell.

Hall, S. 2013. "The Life and Times of Boll Weevil." *Folklife Today* (December 11).

Halpern, S. 1998. "A Fragile Kingdom." *Audubon* 100(2): 36–45.

Hamilton, C.C. 1934. "Combating Insect Pests." *New York Times* (July 8), p. X8.

Hammerstrom, R.J. 1958. "Insect Resistance to Insecticides." *Public Health Reports* 73(12): 1126–32.

Hammond, P., Heinrich, S., Hooker, S., and Tyack, P. 2017. "Whale Culture." *Natural History* 125(10): 30–5.

Hanak, M. 2007. "Afterword." In *Art in Action: Nature, Creativity and our Collective Future*, 170–1. San Rafael, CA: Earthware Editions.

Handy, W.C. 1941. *Father of the Blues*. New York: Macmillan.

Harari, Y.N. 2014. *Sapiens: A Brief History of Humankind*. New York: Harper.

Harbo, J.R. and Rinderer, T.E. 1980. "Breeding and Genetics of Honey Bees." *Beekeeping in the United States Agriculture Handbook Number 335*. United States Department of Agriculture.

Harris, T. 1988. *The Silence of the Lambs*. New York: St. Martin's Press, Macmillan Publishers.

Hatcher, P.E. and Battey, N. 2011. *Biological Diversity: Exploiters Exploited*. Oxford: Wiley-Blackwell.

Hayman, E. 2016. "Art and the Phenomenological Approach to Promoting Environmentalism." *Antennae: The Journal of Nature in Visual Culture* 35: 7–22.

Heard, G. 1950. *Morals Since 1900*. New York: Harper & Brothers.

Hemingway, E. 1925. "A Clean, Well-Lighted Place." *The Short Stories of Ernest Hemingway*. New York: Charles Scribner's Sons.
Hesse-Honegger, C. 1996. *The Future's Mirror*. Newcastle-upon-Tyne, UK: Locus+.
Hesse-Honegger, C. 2001. *HETEROPTERA: The Beautiful and the Other, or Images of a Mutating World*. Zurich, Switzerland: Scalo.
Hesse-Honegger, C. 2008. *Art on Silk*. Zurich, Switzerland: Scheidegger & Spiess.
Hesse-Honegger, C. n.d. *Field Study in Areas of South Vietnam with Agent-Orange Contamination (2004–2005)*. Zurich, Switzerland: Self-published.
Hesse-Honegger, C. and Wallimann, P. 2008. "Malformation of True Bug (Heteroptera): A Phenotype Field Study on the Possible Influence of Artificial Low-Level Radioactivity." *Chemistry & Biodiversity* 5: 499–539.
Higdon, H. 1970. "Well if Not DDT, Then What?" *New York Times* (January 11), p. 220.
Hilbert, M. and Lopez, P. 2011. "The World's Technological Capacity to Store, Communicate, and Compute Information." *Science* (February 10). DOI: 10.1126/science.1200970.
Hitchcock s'explique. 1965. Interview with Alfred Hitchcock on *La Mort aux Trousses* (*North by Northwest*). Available online: https://www.youtube.com/watch?v=GgAPAjRl66w (accessed January 19, 2019).
Hoenes-Stiftung & Dathe, S., eds. 2012. *Jäger und Gejagte (Hunters and Hunted)*. Stuttgart, Germany: Biberacher Verlagsdruckerei.
Holland, J.S. 2013. "The Plight of the Honeybee." *National Geographic* (May 10).
Hollenbach, E. de. 1980. "El mundo animal en el folklore de los triques de Copala." Mexico, DF: *Tlalocan* 8: 437–90. Available online: https://revistas-filologicas.unam.mx/tlalocan/index.php/tl/article/view/237 (accessed May 9, 2017).
Honeybee Genome Sequencing Consortium. 2006. "Insights into Social Insects from the Genome of the Honeybee *Apis mellifera*." *Nature* 443: 931–49. DOI: 10.1038/nature05260.
Hoskins, W.M., Borden, A.D., and Michelbacher, A.E. 1939. "Recommendations for a More Discriminating use of Insecticides." *Proceedings of the 6th Pacific Science Congress of the Pacific Science Association* 5: 119–23.
Howard, L.O. 1921. "a) On Some Presidential Addresses: b) The War Against Insects." *Science* LIV(049): 641–51.
Howard, L.O. 1925. "The Needs of the World as to Entomology." *Annals of the Entomological Society of America* 28(1): 1–21.
Howard, L.O. 1926. "The Insects Intensify their War on Man." *New York Times* (October 10), p. XX3.
Howard, L.O. 1929. "Man's Rival in a Struggle to Survive." *New York Times* (January 20), p. SM84.
Hughes, J.H. 1961. "Entomological Contributions to Human Welfare through the Public Health Service." *Bulletin of the Entomological Society of America* 7(2): 71–5.
Hughes, T.P. 1989. *American Genesis: A Century of Invention and Technological Enthusiasm, 1870–1970*. London: Penguin Books.
Hunter, C. and Aloi, G. 2007. "An Interview with Catherine Chalmers." *Antennae: The Journal of Nature in Visual Culture* 3B: 28–30.
INALI (Instituto Nacional de Lenguas Indígenas). 2008. *Catálogo de las Lenguas Indígenas Nacionales Variantes Lingüísticas de México con us Autodenominaciones y Refereancias Geoestadísticas*. Diario Oficial. Mexico, DF. Available online: www.inali.gob.mx/pdf/CLIN_completo.pdf (accessed May 10, 2017).

Jabr, F. 2013. "The Mind-Boggling Math of Migratory Beekeeping." *Scientific American*. Available online: https://www.scientificamerican.com/article/migratory-beekeeping-mind-boggling-math/ (accessed July 16, 2017).

Jacobsen, R. 2008. *Fruitless Fall: The Collapse of the Honey Bee and the Coming Agricultural Crisis*. New York: Bloomsbury.

Johnson, K. 2010. "With Scientists, Soldiers Solve a Bee Mystery." *New York Times* (October 7), p. A1.

Johnson, R. 2010. "Honey Bee Colony Collapse Disorder." Congressional Research Service, 7-5700, RL33938.

Jundt, T. 2014. *Greening the Red, White, and Blue: The Bomb, Big Business, and Consumer Resistance in Postwar America*. New York: Oxford University Press.

Junger, R. 2002. "E.I. DuPont de Nemours & Company." *The Advertising Age Encyclopedia of Advertising*. Chicago, IL: Fitzroy Dearborn Publisher.

Kaempffert, W. 1945. "Benefits and Dangers of DDT." *New York Times* (September 2), p. 59.

Kaempffert, W. 1947. "Disease-Carrying Insects Killed by Poisoning Blood of Animal Hosts with Chemicals." *New York Times* (December 21), p. E9.

Kafka, F. 1961. *The Metamorphosis*. In *The Penal Colony: Stories and Short Pieces*. New York: Schocken Books.

Kaplan, J. 1966. *Mr. Clemens and Mark Twain: A Biography*. New York: Simon & Schuster.

Karl, F.R. 1991. *Franz Kafka: Representative Man*. New York: Tickner & Fields.

Kelly, F.F. 1931. "The Sinister Insect." *New York Times* (September 27), p. BR11.

Kelly, F.F. 1933. "Insect Enemies Within Our Gates." *New York Times* (March 26), p. BR9.

Kemp, M. 1998. "Hesse-Honegger's Hand-Work." *Nature* 392: 555.

Kidd, S.M. 2002. *The Secret Life of Bees*. New York: Viking Penguin.

King, E.A. 2010. "The Landscape in Art: Nature in the Crosshairs of an Age-Old Debate." *ARTES Magazine*. Available online: http://www.artesmagazine.com/?p=4744 (accessed May 10, 2023).

Kinkela, D. 2011. *DDT and the American Century: Global Health, Environmental Politics, and the Pesticide That Changed the World*. Chapel Hill, NC: University of North Carolina Press.

Klein, B.A. 2013. *Standing on the Shoulders of Wee Giants. Introduction to ECLOSION: A Juried Group Exhibition of Insect-Inspired Art*. Art.Science.Gallery. Austin, TX, USA.

Klein, B.A. 2015. *Encaustics: Repurposing the Architecture of Insects. Introduction to The Buzz Stops Here: An Exhibition of Encaustic Artwork about the Science and Conservation of Bees*. Art.Science.Gallery. Austin, TX, USA.

Klein, B.A. and Brosius, T. 2022. "Insects in Art During an Age of Environmental Turmoil." *Insects* 13: 448. Available online: https://doi.org/10.3390/insects13050448 (accessed May 10, 2023).

Knipling, E.F. 1963. "A New Era in Pest Control: The Sterility Principle." *Agricultural Science Review* 1(1): 2–12.

Knipling, E.F. 1964. "The Sterility Principle of Insect Population Control." *12th International Congress of Entomology*. London, UK.

Knipling, E.F., Bushland, R.C., Babers, F.H., Culpepper, G.H., and Raun, E.S. 1948. "Evaluation of Selected Insecticides and Drugs as Chemotherapeutic Agents against External Bloodsucking Parasites." *The Journal of Parasitology* 34(1): 55–70.

Koerth-Baker, M. 2013. "When Mutant Mosquitoes Attack." *New York Times Magazine* (February 24), pp. 18–20.
Kogan, M. 1998. "Integrated Pest Management: Historical Perspectives and Contemporary Developments." *Annual Review of Entomology* 43(1): 243–70.
Kohn, H. 1971. "Nationalism and Internationalism." In W.W. Wagar (ed.), *History and the Idea of Mankind*, 29, 132. Albuquerque: University of New Mexico Press.
Kohn, H. 1989. "Fields of Dreams: Old Farming Is New Again." *Rolling Stone* (October 5).
Kolbert, E. 2007. "STUNG: Where Have All the Bees Gone?" *The New Yorker* (August 7).
Körblein, A., Hesse-Honegger C. 2018. "Morphological Abnormalities in True Bugs (Heteroptera) near Swiss Nuclear Power Stations." *Chemistry & Biodiversity* 15: e1800099.
Kranzberg, M. 1971. "Science, Technology and the Unity of Mankind." In W.W. Wagar (ed.), *History and the Idea of Mankind*, 135–58. Albuquerque: University of New Mexico Press.
Kritsky, G. and Cherry, R. 2000. *Insect Mythology*. San Jose, CA: Writers Club Press.
Kuper, P. 2015. *Ruins*. London: SelfMadeHero.
Lamas, G., ed. 2004. *Atlas of Neotropical Lepidopera. Checklist: Part 4A (Hesperioidea-Papilionoidea)*. Gainesville, FL: Association for Tropical Lepidoptera, Scientific Publishers.
Lane, J. 1993. "Overwintering Monarch Butterflies in California: Past and Present." In S.B. Malcom and M. Zalucki (eds.), *Biology and Conservation of the Monarch Butterfly*, 335–44. Los Angeles, CA; Natural History Museum of Los Angeles County, Science Series No. 38.
Larmour, D.H.J. 2000. "Pythagoras and the Butterfly: Nabokov's Ovidian Metamorphosis." In P.I. Barta (ed.), *Metamorphosis in Russian Modernism*, 62. Budapest, Hungary: Central European University Press.
Lauren, P.G. 1988. *Power and Prejudice: The Politics and Diplomacy of Racial Discrimination*. Boulder, CO: Westview Press.
Laurence, W.L. 1948. "Means to Control of Insects Shown." *New York Times* (May 13), p. 27.
Laurence, W.L. 1958. "Insect Hormone Suggest a Key to the Aging Process and the Way to Control It." *New York Times* (March 9), p. E9.
Leadbeater, L.W. 1986. "Aristophanes and Kafka: The Dung Beetle Connection." *Studies in Short Fiction* 23(2): 169–77.
Lee, J.M. 1962. "'*Silent Spring*' is Now Noisy Summer." *New York Times* (July 3), p. 21.
Leskosky, R.J. 2007. "The Real Millennium Bug: Giant Arthropod Films in the 21st Century." *Antennae: The Journal of Nature in Visual Culture* 3A: 8–12.
Levine, L.W. 1977. *Black Culture and Black Consciousness: Afro-American Folk Thought from Slavery to Freedom*. Oxford: Oxford University Press.
Linsenmaier, W. 1972. *Insects of the World*. New York: McGraw-Hill.
Lockwood, J.A. 2009. *Six-legged Soldiers: Using Insects as Weapons of War*. New York: Oxford University Press.
Lockwood, J.A. 2013. *The Infested Mind, Why Humans Fear, Loathe, and Love Insects*. Oxford: Oxford University Press.
Lord, J.C. 2005. "From Metchnikoff to Monsanto and Beyond: The Path of Microbial Control." *Journal of Invertebrate Pathology* 89: 19–29.

Losey, J.E. and Vaughan, M. 2006. "The Economic Value of Ecological Services Provided by Insects." *Bioscience* 56: 311–23.
Lyle, C. 1947. "Advancements and Possibilities in Pest Eradication." *Journal of Economic Entomology* 40(1): 1–8.
MacDonald, M. 2004. "Healing Garden." In J.K. Grande (ed.), *Art Nature Dialogues: Interviews with Environmental Artists*. New York: State University of New York Press.
MacKenzie, D. 1999. "Run, Radish, Run." *New Scientist* (December): 36–9.
Malcolm, S.B. 1987. "Monarch Butterfly Migration in North America: Controversy and Conservation." *Trends in Ecology and Evolution* 2(5): 135–8.
Malcolm, S.B. and Zalucki, M., eds. 1993. *Biology and Conservation of the Monarch Butterfly*. Los Angeles, CA: Natural History Museum of Los Angeles County, Science Series No. 38.
Manos-Jones, M. 2000. *The Spirit of Butterflies: Myth, Magic, and Art*. New York: Harry N. Abrams.
Markley, O.W. and Harman, W.W. 1982. *Changing Images of Man*. New York: Pergamon.
Marley, C. 2008. *Pheromone: The Insect Artwork of Christopher Marley*. Petaluma, CA: Pomegranate Communications.
Marquis, D. 1950. *The Lives and Times of Archy and Mehitabel*. Introduction by E.B. White. New York: Doubleday & Company.
Marquis, D. 1961. *archy and mehitabel*. New York: Dolphin Books. (First published 1927.)
Marquis, D. 1996. *archyology: The Long Lost Tales of Archy and Mehitabel*. Hanover, MA: University Press of New England.
Marquis, D. 2006. *The Annotated Archy and Mehitabel*. New York: Penguin.
Matilsky, B.C. 2018. *Endangered Species: Artists on the Front Line of Biodiversity*. Bellingham, WA: Whatcom Museum.
Maza, R. de la. 1987. *Mariposas Mexicanas*. Mexico, DF: Fondo de Cultura Económica, S.A. De C.V.
McGeoch, M.A. 1998. "The Selection, Testing and Application of Terrestrial Insects as Bioindicators." *Biological Reviews* 73(2): 181–201.
McLamb, E. 2011. "The Ecological Impact of the Industrial Revolution." *Ecology Global Network*. Available online: https://ecologyprime.com/2022/12/16/the-ecological-impact-of-the-industrial-revolution/ (accessed 16 July 2017).
McMannis, K. 1998. *A Guide to Zuni Fetishes & Carvings*. Expanded one vol. edn. Tucson, AZ: Rio Nuevo Publishers.
McMannis, K. 2010. *Zuni Fetishes & Carvings*, 2nd edn. Tucson, AZ: Rio Nuevo Publishers.
McMullan, J. 1983. "Zapotec Tapestries: Colorful, Intriguing Show." *State Times* (Baton Rouge), Trend, 1-C.
McWilliams, J.E. 2008. *American Pests: The Losing War on Insects from Colonial Times to DDT*. New York: Columbia University Press.
Michener, C.D. 2007. *Bees of the World*. Baltimore, MD: Johns Hopkins University Press.
Millennium Ecosystem Assessment (Program). 2005. *Ecosystems and Human Well-Being*. Washington, DC: Island Press.
Mills, C. 2017. "Cheerios Pulls Mascot 'Buzz' in Environmental Protest." *U.S. News and World Report*. Available online: https://www.usnews.com/news/national-news/

articles/2017-03-16/cheerios-pulls-honey-nut-cheerios-mascot-in-environmental-protest (accessed May 10, 2023).
Mississippi Blues Commission. 2003. "W.C. Handy Encounters the Blues." Available online: http://www.msbluestrail.org/blues-trail-markers/w-c-handy (accessed May 23, 2018).
Modern Agriculture. 2017. "The Roaming Life of a Commercial Beekeeper." https://modernag.org/biodiversity/commercial-beekeeping-in-agriculture/ (accessed July 16, 2017).
Monarch Biosphere Reserve. 2017. Available online: https://en.wikipedia.org/wiki/Monarch_Butterfly_Biosphere_Reserve (accessed November 6, 2017).
Morris, W. Jr. 1987. *Living Maya*. New York: Harry N. Abrams, Inc.
Muir, J. 1911. My First Summer in the Sierra. San Francisco: Sierra Club Books. 1988 edition.
Naess, A. 1980. *Ecology Community and Lifestyle*. Cambridge: Cambridge University Press.
National Research Council, Committee on the Status of Pollinators in North America. 2007. *Status of Pollinators in North America*. National Academies Press. Available online: https://nap.nationalacademies.org/catalog/11761/status-of-pollinators-in-north-america (accessed May 18, 2023).
Natural World Museum. 2007. *Art in Action: Nature, Creativity and our Collective Future*. San Rafael, CA: Earth Aware Editions.
Nazari, V. 2014. "Chasing Butterflies in Medieval Europe." *Journal of the Lepidopterists' Society* 68(4): 223–31.
Nazari, V. and Evans, L. 2015. "Butterflies of Ancient Egypt." *Journal of the Lepidopterists' Society* 69(4): 242–67.
Nelson, G.C., ed. 2001. *Genetically Modified Organisms in Agriculture: Economics and Politics*. London: Academic Press.
Nietzsche, F. 1927. *The Philosophy of Nietzsche*. New York: Modern Library.
Nisbet, E.K., Zelenski, J.M., and Murphy, S.A. 2011. "Happiness is in our Nature: Exploring Nature Relatedness as a Contributor to Subjective Well-Being." *Journal of Happiness Studies* 12(2): 303–22.
The Norton Anthology of Modern Poetry. 1973. New York: W.W. Norton & Company.
Oberhauser, K., Nail, K., and Altizer, S., eds. 2015. *Monarchs in a Changing World: Biology and Conservation of an Iconic Butterfly*. Princeton, NJ: Cornell University Press.
Oertel, E. 1980. "History of Beekeeping in the United States." *Agriculture Handbook 335*, United States Department of Agriculture.
Office of Technological Assessment. 1995. *Biologically Based Technologies for Pest Control*. US Congress, OTA-ENV-636. Washington, DC: US Printing Office, September.
Olsson, T.C. 2017. *Agrarian Crossings: Reformers and the Remaking of the US and Mexican Countryside*. Princeton, NJ: Princeton University Press.
Opala, J.A. 1987. *The Gullah: Rice, Slavery, and the Sierra Leon-American Connection*. Washington, DC: United States Information Agency.
O'Sullivan, R. 2015. *American Organic: A Cultural History of Farming, Gardening, Shopping, and Eating*. Lawrence: University of Kansas Press.
O'Toole, T. 1966. "A Synthetic Hormone Kills Insects by Keeping them Immature." *New York Times* (April 3), p. 30.
Oxford English Dictionary. 1989. Oxford: Oxford University Press.

Pacquement, A. and Williams, T. 2003. "Environmental Art." Oxford Art Online. Available online: https://doi.org/10.1093/gao/9781884446054.article.T026392 (published online 2003, updated and revised May 25, 2016).
Pedigo, L.P. and Rice, M.E. 2014. *Entomology and Pest Management*, 6th edn. Long Grove, IL: Waveland Press, Inc.
Pell, R.W., Kutil, E., and Turpin, E. 2015. "PostNatural Histories." In H. Davis and E. Turpin, *Art in the Anthropocene: Encounters Among Aesthetics, Politics, Environments and Epistemologies*, 299–316. London, UK: Open Humanities Press.
Perkins, J. 1982. *Insects, Experts, and the Insecticide Crisis: The Quest for New Pest Control Management*. New York: Plenum Press.
Perkins, J. 1983. "Insects, Food, and Hunger: The Paradox of Plenty for U.S. Entomology, 1920–1970." *Environmental Review* 7(1): 71–96.
Peshin, R.P. and Zhang, W. 2014. "Integrated Pest Management and Pesticide Use." In D. Pimentel and R. Peshin (eds.), *Integrated Pest Management: Pesticide Problems*, vol. 3, 1–46. New York: Springer.
Petersen, C. 1976. *Maya of Guatemala: Life and Dress*. Guatemala City, GU: Ixebel Textile Museum.
Peterson, R.K.D. 2007. "Charley Patton and His Mississippi Boweavil Blues." *American Entomologist* 53(3): 142–4.
Peterson, R.K.D.2013. "Mosquito Moan." *American Entomologist* 59: 110–12.
Peterson, R.K.D. 2014. "Mean Old Bedbug Blues." *American Entomologist* 60: 241–3.
Peterson, R.K.D., Higley, L.G., and Pedigo, L.P. 2018. "Whatever Happened to IPM?" *American Entomologist* 64: 146–50.
Pollack, A. 2003. "U.S. Approves Type of Corn that May Cut Pesticide Use." *New York Times* (February 26), p. C10.
Pollack, A. 2014. "Genetic Weapon Against Insects Raises Hope and Fear in Farming." *New York Times* (January 27), p. A1.
Pollack, A. 2016. "New Weapon to Fight Zika: The Mosquito." *New York Times* (January 31), p. A1.
Pollan, M. 2002. *The Botany of Desire: A Plant's-Eye View of the World*. New York: Random House.
Pollan, M. 2006. *The Omnivore's Dilemma: A Natural History of Four Meals*. New York: Penguin.
Population Reference Bureau. 2016. "Human Population: Population Growth." Available online: https://www.prb.org/wp-content/uploads/2016/08/prb-wpds2016-web-2016-1.pdf (accessed July 18, 2017).
Price, A. 1983. "Tapestries of Time." *Sunday Advocate* (Baton Rouge, LA) (June 5), pp. 14–15, cover story.
Prum, R. 2017. *The Evolution of Beauty: How Darwin's Forgotten Theory of Mate Choice Shapes the Animal World—and Us*. New York: Doubleday.
Pyle, R.M. 2001. "Las Monarcas: Butterflies on Thin Ice." *Orion* 20(2): 16–25.
Reba, M., Reitsma, F., and Seto, K.C. 2016. "Spatializing 6,000 Years of Global Urbanization from 3700 BC to AD 2000." *Scientific Data* 3: 160034. DOI: 10.1038/adata.2016.34.
Rees, L. 2017. "Japanese Racism Toward the Chinese." Interview with Akira Iriye. WW2History.com. Available online: http://ww2history.com/experts/Akira_Iriye/Japanese_racism_towards_the_Chinese (accessed January 13, 2017).
Richardson, H.D. 1847. *The Pests of the Farm; With Instructions for Their Extirpation: A Manual of Directions for the Destruction of Vermin*. London: Dublin.

Riley, C.V. 1876. *Potato Pests: Being an Illustrated Account of the Colorado Potato-beetle and Other Insects Foes of the Potato in North America*. New York: Orange Judd Company.

Robertson, T. 2012. *The Malthusian Moment: Global Population Growth and the Birth of American*. New Brunswick, NJ: Rutgers University Press.

Rodee, M. and Ostler, J. 1995. *The Fetish Carvers of Zuni*. Albuquerque, NM: The Maxwell Museum of Anthropology, The University of New Mexico, and Zuni, NM: The Pueblo of Zuni Arts and Crafts.

Ross, G.N. 1964a. "Life History Studies on Mexican Butterflies. II. Early Stages of *Anatole rosssi*, a New Myrmecophilous Metalmark." *Journal of Research on the Lepidoptera* 3(2): 81–94.

Ross, G.N. 1964b. "Life-History Studies on Mexican Butterflies. IV. The Ecology and Ethology of *Anatole rossi*, a Myrmecophilous Metalmark (Lepidoptera: Riodinidae)." *Annals of the Entomological Society of America* 59(5): 985–1004.

Ross, G.N. 1975–7. "An Ecological Study of the Butterflies of the Sierra de Tuxtla in Veracruz, Mexico." *Journal of Research on the Lepidoptera* 14(2): 103–24; 14(3): 169–88; 14(4): 233–52; 15(1): 41–60; 15(2): 109–28; 15(3): 185–200; 15(4): 225–40; 16(1): 87–130.

Ross, G.N. 1985. "The Case of the Vanishing Caterpillar." *Natural History* 94(11): 48–55.

Ross, G.N. 1986. "The Bug in the Rug." *Natural History* 95(3): 66–73.

Ross, G.N. 1987. "A Blue Future for Mexican Indigo." *Américas* 39(4): 40–6.

Ross, G.N. 1988. "Threads of Tradition." *Américas* 40(4): 16–21.

Ross, G.N. 1994. "Winged Victory." *Wildlife Conservation* 97(4): 60–7.

Ross, G.N. 1995. *Everything You Ever Wanted To Know About Butterflies: 100+ Questions and Answers By Gary Noel Ross, Ph.D.* Baton Rouge, LA: Batts Printing.

Ross, G.N. 1996. "Butterfly Houses: Magic Gardens of the 1990s." *American Butterflies* 4(3): 14–21.

Ross, G.N. 1997a. "Goddesses Fly Again: Butterfly Images in Mexican Myth and Textiles." *American Butterflies* 5(4): 14–18.

Ross, G.N. 1997b. "*Lemonias rossi*." In E. Soriano, R. Dirzo, and R. Vogt (eds.), *Historia natural de Los Tuxtlas*. Parte IV. *Invertebrados: Insecta*, 323–8. Mexico, DF: Universidad Nacional Autónomia de México.

Ross, G.N. 2001. "Monarch Section." *Louisiana Wildlife Federation* 29(4): 13–40.

Ross, G.N. 2009. "Exhibit of Zuni Butterfly Art at McGuire Enter Extended Through 2009." *Notes. Association for Tropical Lepidoptera* June: 1–2.

Ross, G.N. 2010a. "Caterpillars, Ants and Populuca (sic) Indians: An Adventure in Remote Mexico." *News of the Lepidopterists' Society* 52(1): 34–41, 44.

Ross, G.N. 2010b. "Corrections, please!" In "The Mailbag ..., Letters to the Editor." *News of the Lepidopterists' Society* 52(2): 69, 72.

Ross, G.N. 2010c. "Living with the Popoluca: Another Adventure in Remote Mexico." *News of the Lepidopterists' Society* 52(4): 118–27.

Ross, G.N. 2012a. "Threads of Tradition: The Butterfly Motif in Mexico's Zapotec Tapestries." *News of the Lepidopterists' Society* 54(3): 98–103.

Ross, G.N. 2012b. "Zapotec Butterfly Tapestries: A Legacy of Pre-Columbian Mythology." *Southern Lepidopterists' News* 34(4): 197–214.

Ross, G.N. 2015a. "My Experiences with Butterfly Imagery in Contemporary Indigenous Cultures. Part 1: Mesoamerica." *Southern Lepidopterists' News* 37(1): 5–15.

Ross, G.N. 2015b. "Air Dancers in Stone: Zuni Butterfly Fetishes are Messengers of a Benevolent Spirit." *Natural History* 123(3): 12–17, cover.
Ross, G.N. 2015c. "My Experiences with Butterfly Imagery in Contemporary Indigenous Cultures. Part 2: Zuni Tribe of New Mexico." *Southern Lepidopterists' News* 37(2): 67–79.
Ross, G.N. 2018. "Four New Zuni Butterfly Fetishes." *Southern Lepidopterists' News*; 40(1): 11–12.
Rothschild, M. 1991. *Butterfly Cooing Like a Dove*. New York: Doubleday.
Royte, E. 1990. "The Ant Man." *New York Times* (July 22), p. SM17.
Russell III, E.P. 1999. "L. O. Howard Promoted War Metaphors as a Rallying Cry for Economic Entomologists." *American Entomologist* 45(2): 74–8.
Russell, E. 2001. *War and Nature: Fighting Humans and Insects with Chemicals from World War I to Silent Spring*. Cambridge: Cambridge University Press.
Russell, L.M. 1978. "Leland Ossian Howard: A Historical Review." *Annual Review of Entomology* 23: 1–15.
Sams, J. and Carson, D. 1988. *Medicine Cards: The Discovery of Power through the Ways of Animals*. Santa Fe, NM: Bear & Company.
Samways, M.J. 2005. *Insect Diversity Conservation*. Cambridge: Cambridge University Press.
Sánchez-Bayo, F., Goulson, D., Pennacchio, F., Nazzi, F., Goka, K., and Desneux, N. 2016. "Are Bee Diseases Linked to Pesticides? – A Brief Review." *Environment International* 89–90: 7–11.
Sarikonda, C. 2016. "Counting Western Monarchs." *Southern Lepidopterists' News* 38(1): 7–11.
Saturday Night Live (SNL). 1976. "The Killer Bees: Home Invasion." National Broadcasting Company, January 10. Available online: https://www.youtube.com/watch?v=APUWOoUqQho (accessed May 18, 2023).
Sauer, P. 2001. "The Monarch Versus the Global Empire." *Orion* 20(2): 42–8.
Schacker, M. 2008. *A Spring Without Bees: How Colony Collapse Disorder Has Endangered Our Food Supply*. Guilford, CT: Lyons Press.
Schappert, P. 2000. *A World for Butterflies: Their Lives, Behavior and Future*. Toronto, Canada: Key Porter Books Ltd.
Schappert, P. 2004. *The Last Monarch Butterfly*. Buffalo: Firefly Books (U.S.), Inc.
Schlesinger Jr., A. 1969. "Origins of the Cold War." In B.J. Bernstein and A.J. Matusow (eds.), *Twentieth Century America: Recent Interpretations*, 437–58. New York: Harcourt, Brace & World.
Schroeder, P.W. 1969. "The Axis Alliance and Japanese-American Relations, 1941: An Appraisal of the American Policy." In B.J. Bernstein and A.J. Matusow (eds.), *Twentieth Century America: Recent Interpretations*, 169. New York: Harcourt, Brace & World.
Selcraig, B. and Borris, D. 2003. "Dream Weavers." *Smithsonian* 34(8): 72–8.
Semmens, B.X., Semmens, D.J., Thogmartin, W.E., Wiederholt, R., López-Hoffman, L., Diffendorfer, J.E., Pleasants, J.M., Oberhauser, K.S., and Taylor, O.R. 2016. "Quasi-Extinction Risk and Population Targets for the Eastern, Migratory Population of Monarch Butterflies (*Danaus plexippus*)." *Scientific Reports* 6: 23265.
SFU Museum of Archaeology and Ethnology. 2010. "Forensic Entomology or the Use of Insects in Death Investigations." *Investigating Forensics*. Available online: https://www.sfu.ca/~ganderso/forensicentomology.htm (accessed July 16, 2017).

Shannon, D.A. 1974. *20th Century America: The Twenties and Thirties*, vol. 2, 3rd edn. Chicago, IL: Rand McNally.

Shin, G.-W. 2006. *Ethnic Nationalism in Korea: Genealogy, Politics, and Legacy*. Palo Alto, CA: Stanford University Press.

Shiva, V. 1991. *The Violence of the Green Revolution: Third World Agriculture, Ecology, and Politics*. London: Zed Books.

SIL. 2017. Available online: www.sil.org (accessed November 6, 2017).

Smart Museum of Art. 2000. "Ecologies: Mark Dion, Peter Fend, Dan Peterman." University of Chicago, Illinois USA, July 6–August 27, 2000.

Smith, A. 2005. "Blues, Criticism, and the Signifying Trickster." *Popular Music* 24: 179–91.

Smith, E.H. 1989. "The Entomological Society of America: The First Hundred Years, 1889–1989." *Bulletin of the Entomological Society of America* 35(3): 10–32.

Smith, H.S. 1941. "Racial Segregation in Insect Populations and its Significance in Applied Entomology." *Journal of Economic Entomology* 34(1): 1–13.

Smith, S. 2001. *Ecologies: Mark Dion, Peter Fend, Dan Peterman*. Chicago, IL: Smart Museum of Art, University of Chicago.

Smithers, P. 2008. *Insects and Art: A Contemporary and Historical Perspective*. Plymouth, UK: Peninsula Arts, University of Plymouth.

Sorell, T. 1991. *Scientism: Philosophy and the Infatuation with Science*. New York: Routledge.

Soriano, E., Dirzo, R., and Vogt, R. 1997. *Historia Natural de los Tuxtlas*. Mexico, DF: Universidad Nacional Autónoma de México, Instituto de Biología, Apartado Postal 70-143, Instituto de Ecología, Apartago Postal 70-275, 94510.

Sparks, T.C. and Nauen, R. 2015. "IRAC: Mode of Classification and Insecticide Resistance Management." *Pesticide Biochemistry and Physiology* 121: 122–8.

Stackman, D.C. 2005. *Orange Empire: California and the Fruits of Eden*. Berkeley, CA: University of California Press.

Stanton, A. 1999. *Zapotec Weavers of Teotitlan*. Santa Fe, NM: Museum of New Mexico Press.

Tauger, M.B. 2011. *Agriculture in World History*. New York: Routledge.

Thogmartin, W.E., López-Hoffman, L., Rohweder, J., Diffendorfer, J., Drum, R., Semmens, D., Black, S., Caldwell, I., Cotter, D., Drobney, P., Jackson, L.L., Gale, M., Helmers, D., Hilburger, S., Howard, E., Oberhauser, K., Pleasants, J., Semmens, B., Taylor, O., Ward, P., Weltzin J.F., and Wiederholt, R. 2017. "Restoring Monarch Butterfly Habitat in the Midwestern US: 'All Hands On Deck'." *Environmental Research Letters* 12(7).

Thomas, J. 1998. "With Change in Habit, Beetles Shake Up Corn Country." *New York Times* (October 20), p. F3.

Thornes, J.E. 2008. "A Rough Guide to Environmental Art." *Annual Review of Environment and Resources* 33: 391–411.

Tonn, S. 2015. "Termites Are Teaching Architects to Design Super-Efficient Skyscrapers." *Wired Magazine*. Available online: https://www.wired.com/2015/11/building-skyscrapers-like-termite-mounds-could-save-energy/ (accessed July 16, 2017).

Tsutsui, W. 2007. "Looking Straight at *THEM!*: Understanding the Big Bug Movies of the 1950s." *Environmental History* 12: 237–53.

United States Department of Agriculture (USDA). 1952. *Insects: Yearbook of Agriculture*. Washington, DC: United States Department of Agriculture.

United States Department of Agriculture (USDA). 2012. "Agriculture Then and Now." *USDA National Agriculture Library*. Available online: https://archive.org/details/CAT31294574/page/n1/mode/2up (accessed July 20, 2017).

Urquhart, A. and Lavies, B. 1976. "Found at Last: The Monarch's Winter Home." *National Geographic* 150(2): 160–73, cover.

Urquhart, F. 1987. *The Monarch Butterfly: International Traveler*. Chicago, IL: Nelson-Hall.

van der Linden, S., Maibach, E., and Leiserowitz, A. 2015. "Improving Public Engagement With Climate Change." *Perspectives on Psychological Science* 10(6): 758–63.

Vane-Wright, R. 1993. "The Columbus Hypothesis: An Explanation for the Dramatic 19th Century Range Expansion of the Monarch Butterfly." In S.B. Malcolm and M. Zalucki (eds.), *Biology and Conservation of the Monarch Butterfly*, 179–80. Los Angeles, CA; Natural History Museum of Los Angeles County, Science Series No. 38.

Various. 1922. *American Poetry; A Miscellany*. New York: Harcourt, Brace and Company.

Warren, R. 2015. "A Designing Mind." *American Poetry Review* 44(1): 27–30.

Watson, L.R. 1928. "Controlled Mating in Honeybees." *The Quarterly Review of Biology* 3: 377–90.

Webster, B. 1979a. "New Approaches Sought in War on Pests." *New York Times* (June 5), p. C1.

Webster, B. 1979b. "Nice Bugs to the Rescue as Pesticides Lose their Zip." *New York Times* (September 23), p. E9.

Wellhouse, W.H. 1926. *How Insects Live: An Elementary Entomology*. New York: The Macmillan Company.

Whitecotton, J. 1977. *The Zapotecs: Princes, Priests, and Peasants*. Norman, OK: University of Oklahoma Press, The Civilization of the American Indian Series.

Williams, E.H. and Brower, L. 2016. "A Conservation Concern: How Many Monarchs Are There?" *News of the Lepidopterists' Society* 58(2): 90–3.

Wilson, E.O. 2002. *The Future of Life*. New York: Vintage.

Wolkenfeld, S. 1976. "Psychological Disintegration in Kafka's 'A Fratricide' and 'An Old Manuscript.'" *Studies in Short Fiction* 13(1): 25–9.

Wordsworth, W. 1984. *William Wordsworth*. Oxford: Oxford University Press.

World Health Organization. 2014. *A Global Brief on Vector-borne Diseases* (WHO document WHO/DCO/WHD/2014.1).

Worldwatch Institute. 2010. "State of the World: Transforming Cultures from Consumerism to Sustainability." New York: W.W. Norton and Company. Available online: https://archive.org/details/stateofworld20100000unse (accessed July 18, 2017).

Worster, D. 2004, 1979. *Dust Bowl: The Southern Plains in the 1930s*. New York: Oxford University Press.

Yeats, W.B. 1986. *Selected Poems and Three Plays of William Butler Yeats*, 3rd edn. New York: Collier Books.

Yoldas, Pinar. 2015. "Ecosystems of Excess." In H. Davis and E. Turpin, *Art in the Anthropocene: Encounters Among Aesthetics, Politics, Environments and Epistemologies*. London, UK: Open Humanities Press.

ACKNOWLEDGMENTS

Chapter 8

We thank Amey Zhang, Michael Samways, Clair Kovacs, Margaret Morse, and Arno Klein for reading an earlier version of this chapter, to the artists who helped clarify and interpret their motivations when producing art featuring insects, and to all of the artists who gave permission for image use. We produced a companion piece to this chapter for the journal *Insects* (Klein and Brosius 2022), and thank the artists surveyed within.

INDEX